47

Advances in Polymer Science

Fortschritte der Hochpolymeren-Forschung

Synthesis and Degradation Rheology and Extrusion

With Contributions by
A. L. Bhuiyan M. Dröscher W. W. Graessley
E. Neuse

With 37 Figures

Springer-Verlag
Berlin Heidelberg New York 1982

Editors

ISBN-3-540-11774-1 Springer-Verlag Berlin Heidelberg New York
ISBN-0-387-11774-1 Springer-Verlag New York Heidelberg Berlin

Library of Congress Catalog Card Number 61-642

Typesetting and printing: Schwetzinger Verlagsdruckerei. Bookbinding: Brühlsche Universitätsdruckerei, Gießen.
2152/3140 – 543210

Table of Contents

Aromatic Polybenzimidazoles. Syntheses, Properties, and Applications

Eberhard W. Neuse

Department of Chemistry, 1 Jan Smuts Avenue, Johannesburg 2001, Republic of South Africa

Outstanding performance characteristics under unusual and demanding operating conditions are the hallmarks of a class of macromolecular compounds commonly referred to as polyheteroaromatics. The all-aromatic polybenzimidazoles, which rank among the leading representatives of this class, are the topic of the present article. Beginning with the early report of a successful synthesis of poly(1,4-phenylene-5,5′(6′)-bibenzimidazole-2,2′-diyl) from Carl Marvel's laboratory, a 20-years' span of worldwide research activities has until now provided a wealth of synthetic information in the field of aromatic benzimidazole polymers, which is critically surveyed. The problems involved in the conventional high-temperature processing technology are pointed out, and recent preparative developments permitting a two-stage "prepolymer" approach are reviewed. Structure-property relationships are treated in detail, with special attention paid to the problem of high-temperature stability and flame retardancy. A major section of the article presents a discussion of application and performance data in some of the more advanced segments of materials technology.

Advances in Polymer Science 47

Introduction

Polymeric materials composed of aromatic and heteroaromatic ring structures as chain constituents, hereinafter designated as polyheteroaromatics, have in recent years advanced to the forefront of materials technology on account of their valuable engineering properties, such as flame retardance, radiative stability, excellent mechanical and dielectric strength and strength retention over a wide range of temperatures, as well as toughness, chemical inertness, and adhesion. Although it is especially in high-temperature environments, that is, in the regions of 250–300 °C for long-term exposure and of 400–600 °C for short-term exposure, where the polyheteroaromatic engineering materials have proven their superiority, excellent performance at sub-zero temperatures has also been reported for a number of types. The polyheteroaromatics may thus find use as matrix resins in structural reinforced composites for advanced aircraft, aerospace, and electronics designs, as heat- and sound-absorbing materials, structural adhesives, and protective coatings in general technology, as cable and wire insulating compositions in high-performance electrical applications, and, lastly, as fibers, films, and fabrics for use under extreme operating conditions with respect to low and high temperatures, radiative attack, or risk of ignition and combustion.

An outstanding feature of most polyheteroaromatics is their easy processibility in the so-called prepolymer stage. All-aromatic heterocyclic polymers generally possess too high glass transition temperatures to be processible by conventional thermoforming, molding, and casting techniques. They may, however, be processed in a precursor stage, in which they still exhibit sufficient solubility in suitable solvents to permit the use of more or less conventional processing methods involving impregnation, prepreg layup, and lamination. The primary polycondensation product of the precursor stage, the *prepolymer,* is designed so as to contain certain open-chain segments capable of ring closure and aromatization upon heat treatment. The overall approach, utilizing well-established synthesis principles of heterocyclic chemistry, thus comprises two steps. The first step involves the primary preparation of a soluble and processible prepolymer, usually by low-temperature solution polymerization, and its application in the form of lacquers, varnishes, or molding powders. In the second step, the prepolymer, freed from solvent and formed into the final shape in suitable molds, dies or other forming devices, is heated with or without the application of external pressure, whereupon *in situ* cyclocondensation leads to the ultimate, aromatized polyheterocyclic structure. Some instructive illustrations of the concept of two-stage polymer synthesis by ring-forming condensations can be found in the specialized literature[1–7].

The reaction types employed in such two-stage heteroaromatic polymer syntheses generally are borrowed from classical, non-polymeric heterocyclic chemistry, the major differences being that difunctional rather than monofunctional reactants must be used in order to achieve the objective of linear chain formation. Yet some critical limitations in the choice of reaction type must be accepted by the polymer chemist intent on synthesizing heterocyclic macromolecules of high chain length and structural purity. Thus, an extent of reaction higher than 0.99 (corresponding to more than 99% conversion along the propagation path) is required for the number-average degree of polymerization to attain a level (>100) at which the polymer will possess useful engineering properties; this, in turn, implies that the reaction type involved in propagation must be sufficiently unique, i.e., rapid in relation to concurrent side reactions, to limit the extent of such side reactions to a mere fraction of a percent. Moreover, no one side reaction is allowed to yield products capable of participating in the propagation sequence, as this would invariably lead to structural deviations and might even cause branch generation or crosslinking. These stringent requirements eliminate the great majority of conventional heterocycle-forming reaction paths, which, although satisfactory for synthetic purposes in the non-polymeric domain, are accompanied by side reactions to an extent entirely unacceptable to the polymer chemist, and it is largely because of those restrictive requirements that a disproportionately small number of polyheteroaromatic structures has found commercial acceptance. Foremost among these are the polypyromellitimides – the only type of heteroaromatic polymers, in fact, that has reached the stage of large-scale commercialization. Other types, while perhaps even more promising from an application standpoint, have not to this date attained the same level of commercial maturity as a consequence of existing production or processing problems. This includes the benzimidazole-containing polymers, which are the topic of the present article. In the following section a survey will be given of the preparative chemistry of the polybenzimidazoles, including recent developments involving two-stage synthesis techniques. Although the early polybenzimidazole chemistry has been proficiently reviewed many years ago[8)], its more significant aspects will be reiterated in this survey so as to present the subject in a comprehensive and appropriately rounded format. Subsequent sections will provide a review of the more important properties and outstanding technological applications of this eminently promising class of polyheteroaromatics. The discussion will be restricted to polymers having non-aliphatic bridging segments in the backbone (commonly designated as "all-aromatic", even though such non-aromatic groupings as ether, sulfone, or carbonyl may be present as chain constituents), since polybenzimidazoles with aliphatic segments do not in general excel in properties of current technological interest. Also excluded are those polyheteroaromatics in which the benzimidazole system merely constitutes an annelated component in other fused ring structures, as in the polybenzimidazopyrrolones or polybenzimidazobenzophenanthrolines. The polymers discussed in the following all contain benzimidazole segments that are either of the type *A* or of the type *B*. Although, in solution, there is rapid interconversion between tautomeric structures as shown, neither the nomenclature employed nor the structural formulae reproduced in this article do reflect such tautomerism, only the left-hand-side representations being used throughout for reasons of simplicity. In accordance with IUPAC nomenclature rules, assigning the locant 1 to the imino (–NH–) nitrogen atom of the imidazole ring, these representations will be named, respectively, benzimidazole-2,5-diyl and imidazo[4,5-*f*]benzimidazole-2,6-diyl.

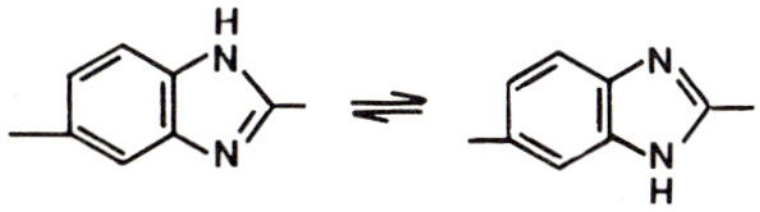

A

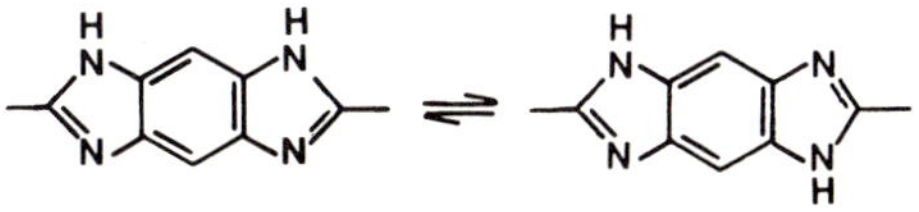

B

1 Syntheses

1.1 Melt and Solid-state Polymerizations

The year 1961 marks the beginning of one of the most productive periods of research in the science of advanced materials. In this year Carl S. Marvel, one of the foremost researchers in the field of heterocyclic polymer chemistry, reported the first synthesis of all-aromatic, high-molecular-mass polybenzimidazoles[9], thus "kicking off", as it were, the challenging game of devising efficient synthetic approaches toward a multitude of fully aromatic polyheterocyclic macromolecules for diversified use in current technology. Marvel's pioneering approach involved the melt polycondensation of 3,3′-diaminobenzidine or 1,2,4,5-tetraaminobenzene with the diphenyl esters of a large number of aromatic dicarboxylic acids (Schemes 1 and 2; Ar = 1,3- or 1,4-phenylene, 4,4′-biphenylene, etc.). The reaction was carried out by heating, under nitrogen, the equimolar mixture of tetraamine and diphenyl ester at temperatures gradually increasing from about 200 to 300 °C. When the initially liquid melt had solidified, heating was continued for a short time under reduced pressure. The pulverized intermediary polymeric solid, when post-heated for several hours at 350–400 °C under a high vacuum, underwent a nearly quantitative conversion to the fully aromatized polybenzimidazole structure. In a similar fashion, benzimidazole polymers were obtained by homopolymerization of monomers that comprised both a phenoxycarbonyl function and an *o*-diaminobenzene moiety, such as phenyl 3,4-diaminobenzoate (Scheme 3). Whereas the products of the first-named two reaction types contained bibenzimidazolediyl or benzodiimidazolediyl groups as chain constituents, each one of these interconnected by an arylene unit, the last-named reaction gave rise to polymer chains made up exclusively of benzimidazolediyl repeating units.

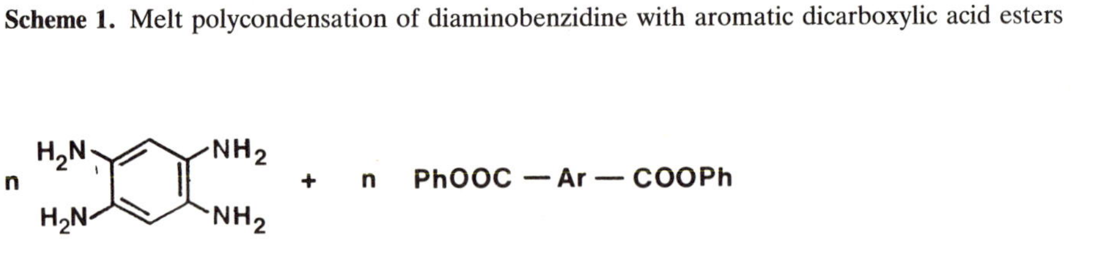

Scheme 1. Melt polycondensation of diaminobenzidine with aromatic dicarboxylic acid esters

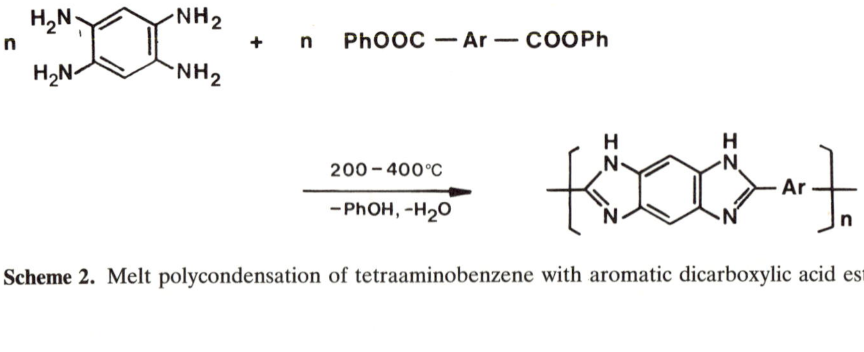

Scheme 2. Melt polycondensation of tetraaminobenzene with aromatic dicarboxylic acid esters

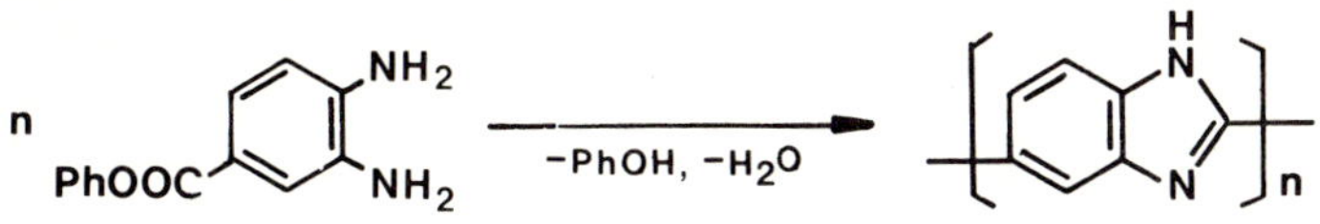

Scheme 3. Homopolycondensation of a diaminobenzoate in the melt

With but minor modifications, Marvel's original procedure, since extended by his own school[10–14] and many others[15–21] (leading references only) to include numerous additional monomer pair combinations, has been in use until now as the principal approach for laboratory synthesis, specialty preparation in adhesive, foam, and fiber applications, and pilot plant development manufacturing. Other monomers employed in these investigations include, *inter alia,* bis(3,4-tetraaminodiphenyl) ether, bis(3,4-tetraaminodiphenyl) sulfone, bis(3,4-tetraaminodiphenyl) ketone, and 1,3-dianilino-4,6-diaminobenzene, as well as a large number of additional bis(phenoxycarbonyl) derivatives of mono-, di-, and trinuclear aromatic and heterocyclic parent compounds. A comprehensive listing of monomers can be found in Levine's review[8], and a number of additional monomer reactants used in more recent years has been cited by Cassidy[7] and by Babé and de Abajo[22] in later surveys of the field. Representative polybenzimidazoles prepared by Marvel's polymerization technique are the types *1–5, 7, 13–17, 19* and *20.*

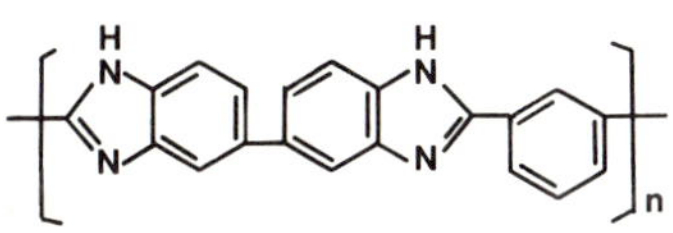

1

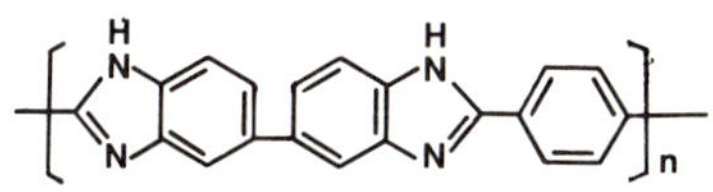

2

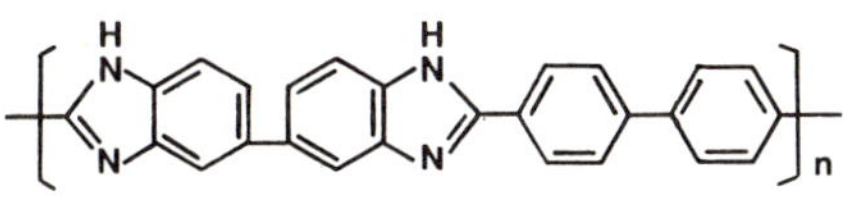

3

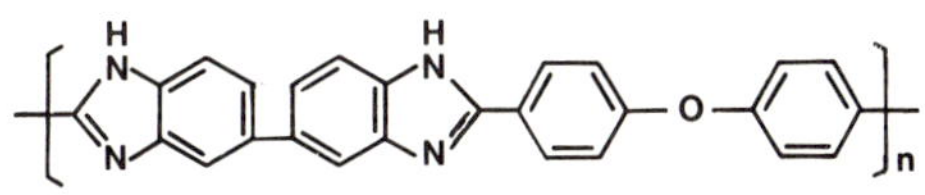

4

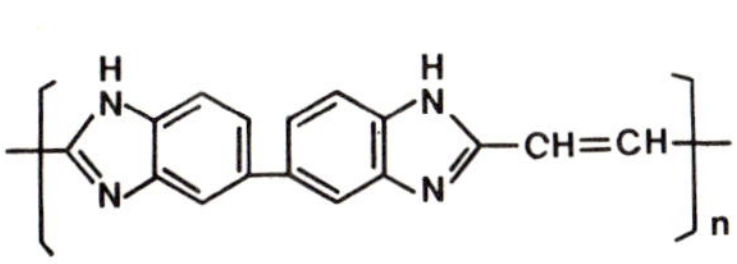

5

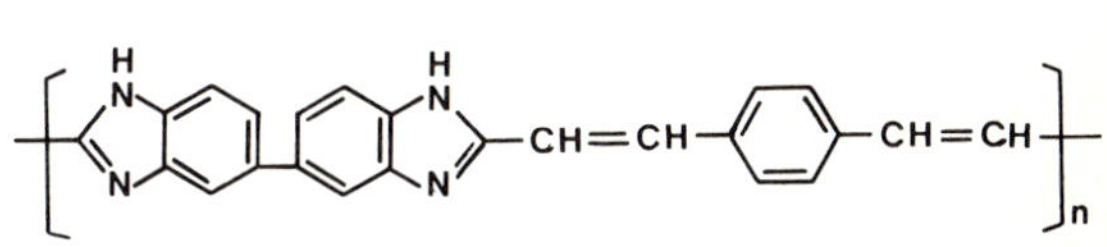

6

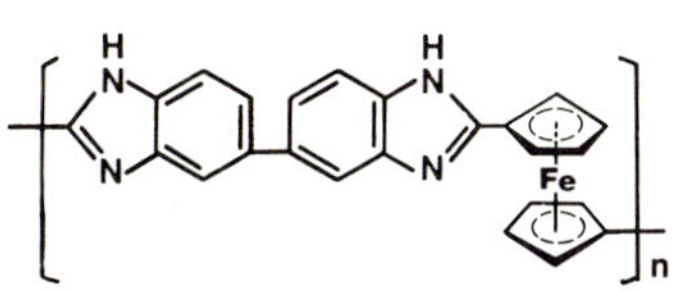

7

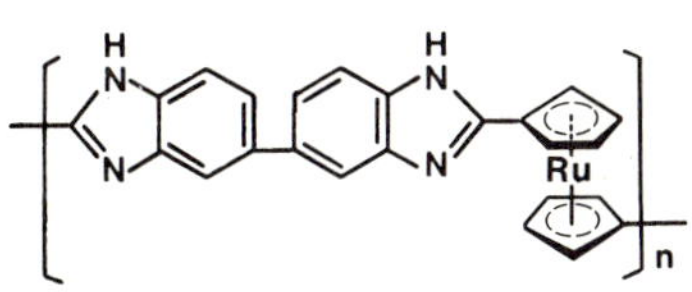

8

Scheme 4. Representative polybenzimidazoles

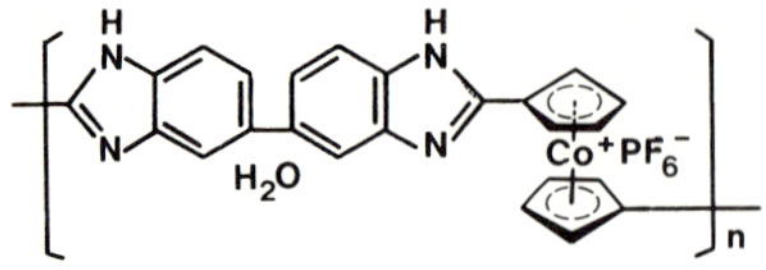

9

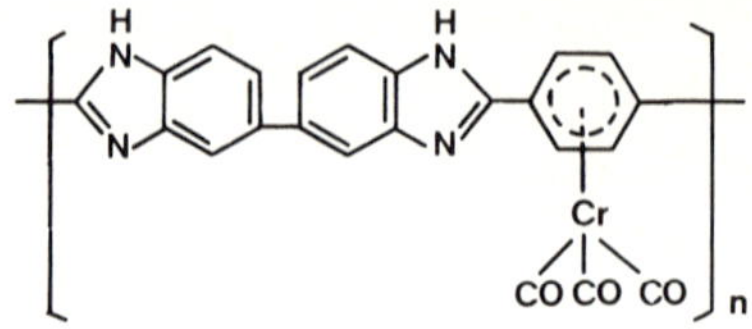

10

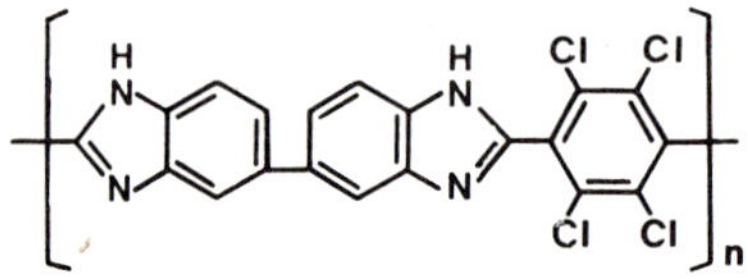

11

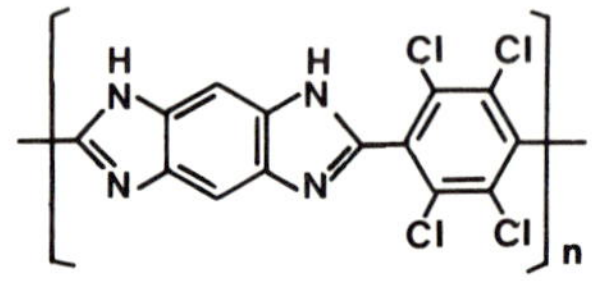

12

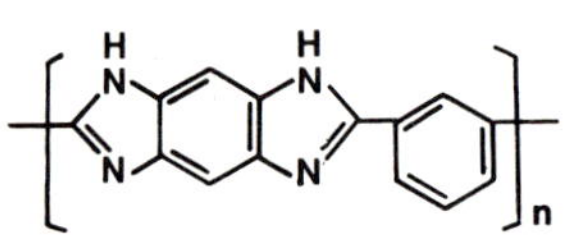

13

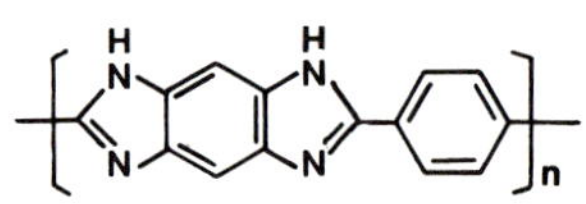

14

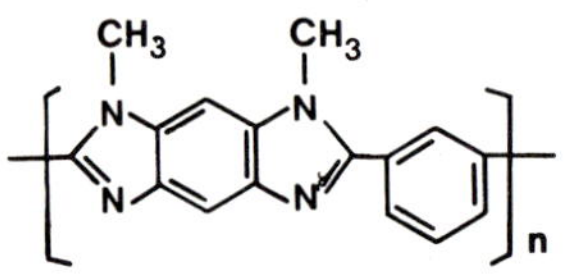

15

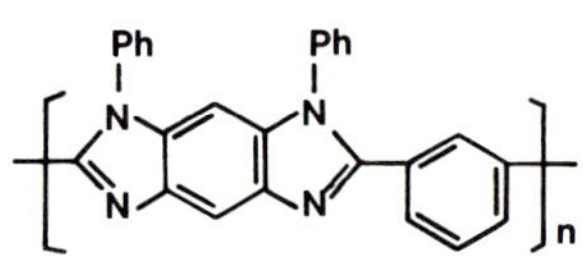

16

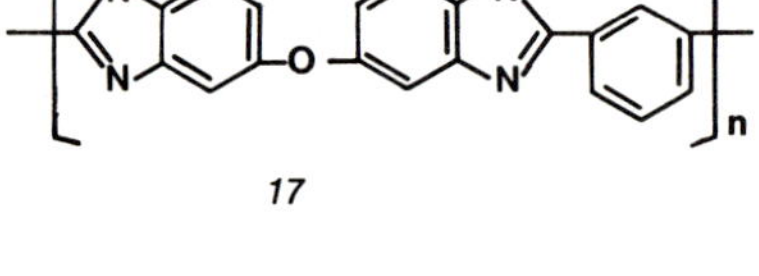

17

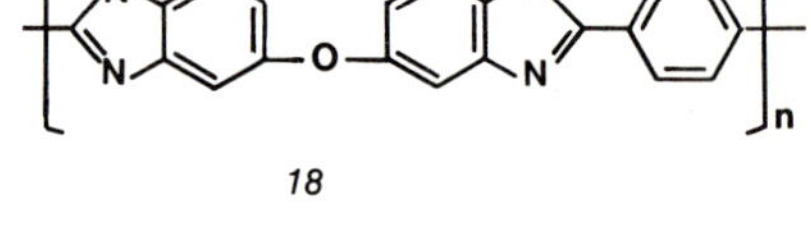

18

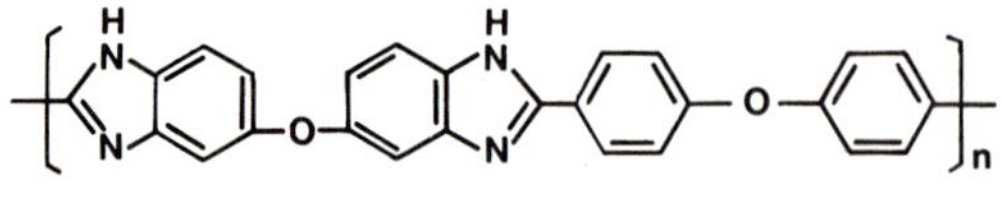

19

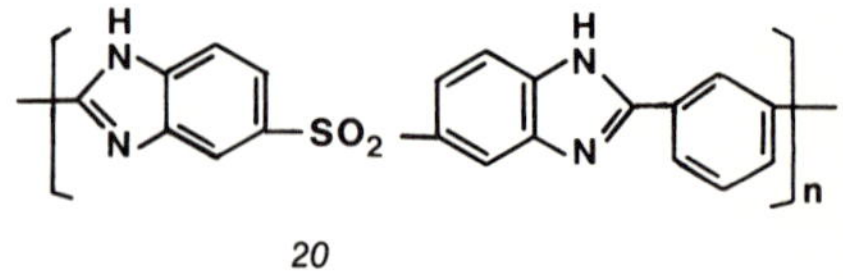

20

Scheme 4 (continued)

The mechanism of the melt condensation process (Schemes 1–3) originally was believed to involve straightforward aminolysis of the phenyl ester in the first step, resulting in poly(aminoamide) formation (segment structure *C*) with loss of phenol. The second step would then represent a simple cyclodehydration to the imidazole system, water

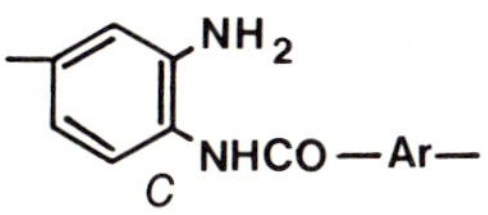

C

being eliminated in this advanced stage of the overall process. However, an elaborate kinetic study conducted more recently[23] provided results contradictory to this premise. By monitoring, as a function of time at two different temperature levels, the consumption of the tetraamine reactant and the evolution of both water and phenol, the authors[23] found that water generation preceeds the expulsion of phenol, the latter being rate determining. The mechanism proposed on the basis of the kinetic results comprises in the first stage the addition of amine nucleophile to the ester carbonyl, followed by the elimination of water. The *C*-phenoxy-substituted azomethine polymer *D* thus generated, probably in equilibrium with the cyclized benzimidazoline tautomer *E* (Scheme 5), eliminates phenol in the second and final stage, thereby aromatizing to the polybenzimidazole end product. These mechanistic findings provide a ready explanation for the porosity observed in ultimate polymer parts manufactured by the melt polycondensation technique. Final-stage phenol evolution will clearly cause a considerably higher extent of pore generation in the ultimate product than would be expected for a process in which the

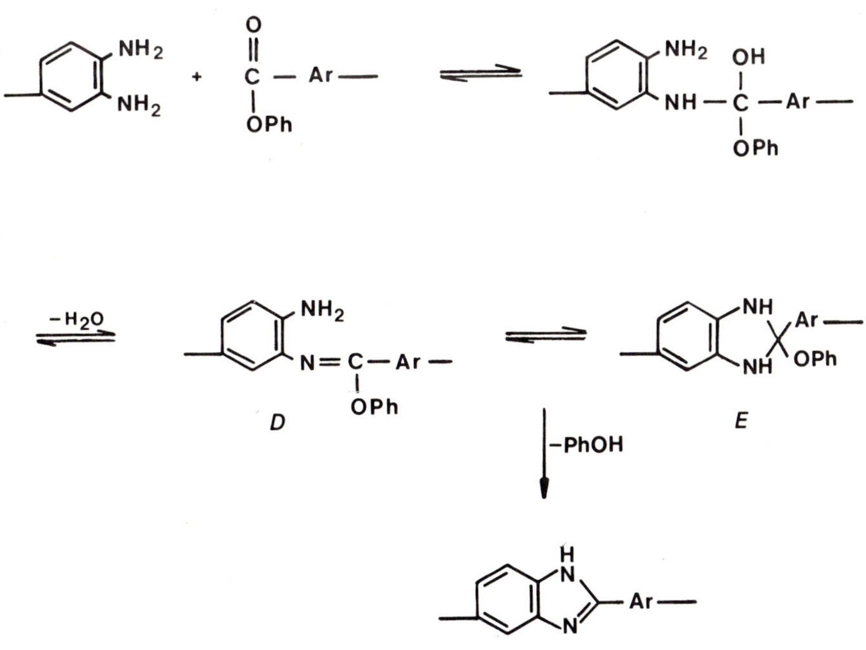

Scheme 5. Mechanism of polybenzimidazole formation by melt polymerization of bis(*o*-diamine) with aromatic bis(phenoxycarbonyl) reactant

bulky phenol molecule is expelled at an early melt stage and the much smaller and more readily diffusing water molecule is the species leaving the (appreciably consolidated) melt system in the final stage.

Although the polybenzimidazoles prepared in these investigations by the melt polycondensation technique by and large have been found to exhibit good to excellent performance characteristics, they do not lend themselves to processing by standard techniques other than spinning and film casting, and in frequent cases have been found to lack satisfactory solubility in non-corrosive solvents. Since the preparative procedure fails to provide well defined, soluble prepolymers to be used for the subsequent and finalizing processing step, one has to resort to cumbersome approaches of employing ill-defined intermediary stages for subsequent hot-melt processing, a feature which has seriously hampered further progress on the application side.

1.2 High- and Low-Temperature Solution Polymerizations

Recognizing the shortcomings of the melt condensation approach, numerous research groups have directed their attention to the task of finding new preparative two-stage, prepolymer-generating procedures. The seemingly obvious answer to the problem would be the low-temperature solution or interfacial polycondensation employing a bis(*o*-diamine) and the acid chloride of a dicarboxylic acid reactant in an effort to intercept and isolate a soluble poly(aminoamide) intermediate (*F*). However, attempts made toward this end remained unsuccessful, as the primary amidation step proved too fast to be selective, causing additional amidation at the *o*-amino group with resultant formation of the segment structure *G*. This side reaction inevitably would lead to branching and,

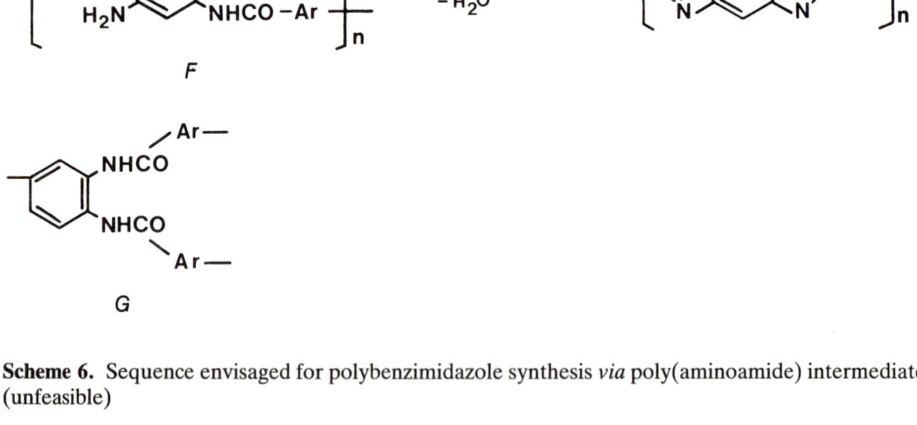

Scheme 6. Sequence envisaged for polybenzimidazole synthesis *via* poly(aminoamide) intermediate (unfeasible)

ultimately, three-dimensional crosslink formation; at the same time, it would prevent "imidazolation", i.e., closure of the five-membered heterocycle. One should add, however, that various approaches have been described more recently in which the problem of multiple amidation and branching is overcome by use of monomers having amino groups of different basicities, thus allowing for a more selective substitution and an enhanced degree of linearity. Replacing tetraaminobenzene (Scheme 6) by 2,3,5,6-tetraaminopyridine, for example, does afford soluble (aminoamide) intermediates that may be employed as prepolymers in a two-stage manufacturing process[24], and similar results are obtained with bis(*o*-anilinoamine)s or bis(*o*-acetamidoamine)s as the tetraamine monomers, *N*-phenyl- or *N*-methyl-substituted benzimidazole polymers being the ultimate products in these latter two cases[25–27]. Use of bis(*o*-nitroaniline)s in place of tetraamines has also been made, and the intermediary poly(*o*-nitroamide)s are subjected to reductive cyclization[28]. In a variation of this theme the polycondensation of 4,4′-dichloro-3,3′-dinitrobenzophenone with 4,4′-(1,4-phenylene dioxy)dianiline, followed by reduction of the nitro groups in the resultant polyamine, gives a poly(*o*-aminoamine). Benzoylation and thermal cyclodehydration of this intermediate affords a 2-phenyl-substituted polybenzimidazole in which the backbone interconnection of the benzimidazole unit occurs at positions 1 and 5 (Scheme 7)[29].

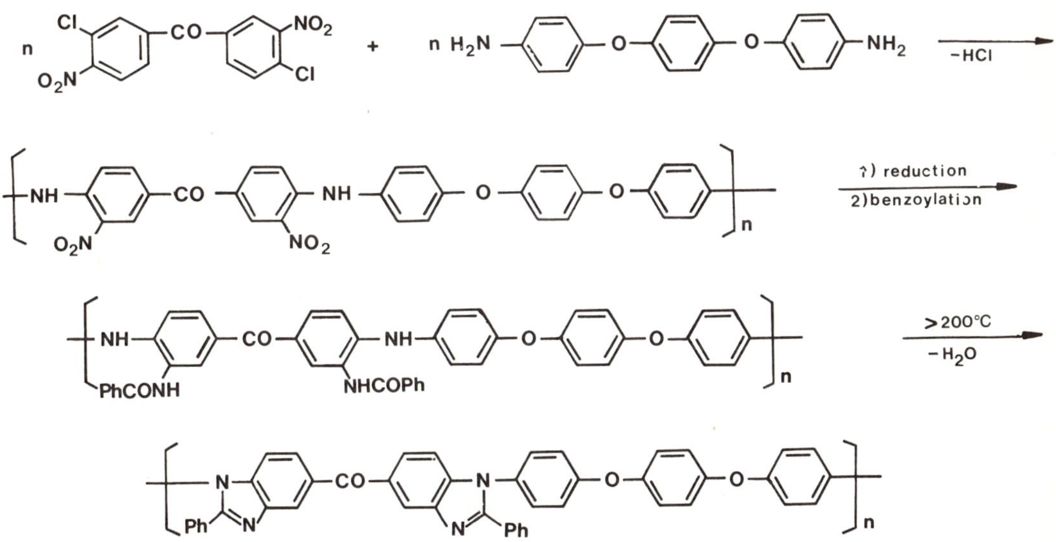

Scheme 7. Synthesis of polymers containing the 2-phenylbenzimidazole-1,5-diyl unit

Numerous communications describe the use of high-temperature solution polymerization techniques. Of the solvents used, poly(phosphoric acid), introduced in 1964 into polybenzimidazole chemistry by Iwakura et al.[30], doubtlessly occupies the first position, and a great many syntheses employing the poly(phosphoric acid) solution technique have since been reported from various polymer laboratories[31–39]. Iwakura's technique offers an advantage insofar as the tetraamine monomers may be employed as the stable and easy-to-handle hydrochlorides in place of the extremely air-sensitive free bases. In addition, the method permits the use of the free dicarboxylic acids as monomers, most of which are too sensitive to survive the high temperatures of the classical process without undergoing major degradative decarboxylation. The reaction is commonly carried out by preheating the tetraamine hydrochloride in degassed poly(phosphoric acid) at 140–150 °C to liberate the base. Upon the addition of the diacid or other easily accessible diacid derivatives, heating is continued for several hours at typically 180–200 °C under nitrogen until cyclization has been completed. The resultant benzimidazole polymers are essentially identical with and possess the same properties as the corresponding products of Marvel's melt condensation process. Among the many novel polybenzimidazoles synthesized in poly(phosphoric acid), one should mention polymers containing an aromatic imide group[38)], the thermostable phenoxathiin[32)] or quinazolinedione[33)] units, the flame-retarding phosphine oxide function[37)], or the cobalticenium cation[34)] in the chain.

While the poly(phosphoric acid) solution polycondensation method presents a convenient means of synthesizing polybenzimidazoles, it is not amenable to prepolymer isolation and processing. It is, therefore, restricted in its usefulness to such cases where the ultimate prodcuts can be utilized, for example, in spinning or film casting. Another drawback of poly(phosphoric acid) solvent is its propensity for promoting the selfcondensation of amines; Marvel's group in fact has shown that 1,2,4,5-tetraaminobenzene and other bis(*o*-diamine)s undergo rather smooth polycondensation when heated *per se* in poly(phosphoric acid), giving highly fused aromatic polymer structures containing dihydrophenazine groupings in the chain[40)]. Furthermore, poly(phosphoric acid) tends to be incorporated into, or strongly adsorbed to, polar polymeric compounds, thereby creating phosphorus contents in the product[8)]; it also has been reported[41)] to attack glass at elevated temperatures, which may lead to incorporation of silica into the polymer, adversely affecting elemental analysis results. Lastly, one must recognize that poly(phosphoric acid) is unsuitable for polymerization of monomers unstable in a hot acidic environment. Thus, for example, the synthesis of the aforementioned cobalticenium-containing polymers[34)] was found to be accompanied by proton-induced cleavage of the metal-ring bond in the complex; ethylated poly(phosphoric acid) proved more satisfactory on account of the appreciably reduced acidity of the medium, giving a polysalt essentially of structure *9*, Scheme 4.

Other solvents used in high-temperature solution polymerizations, the preferred reactants being of the types employed in the melt condensation process, include degassed phenol, veratrol, *N,N*-dimethylaniline, and various amide solvents, such as *N,N*-dimethylacetamide and hexamethylphosphoramide[8, 10, 42)]. By and large, the results were unsatisfactory; while the use of phenol solvent required solid-state heating at high temperatures in the second stage to achieve aromatization, the amide solvents caused complications because of transamidation. It was only by use of sulfolan solvents in Marvel's laboratory[43)] that smooth polymerization of bis(*o*-diamine)s and bis(phenyl ester)s could be accomplished without the side effects observed with poly(phosphoric acid) and the amide media. Linear polybenzimidazoles of high molecular mass were formed readily, provided that the polycondensations were performed at the solvents' boiling temperatures under nitrogen. By using the highly reactive bis(trialkyl ortho-ester)s in lieu of bis(phenyl ester)s, the solvent being a 9 : 1 dimethyl sulfoxide/pyridine mixture, other workers[44)] were able to demonstrate more recently that the same results can be achieved (Scheme 8) within an hour or less at temperatures as low as 100 °C. Although the reaction, in common with other, earlier reported solution polymerizations, proceeds

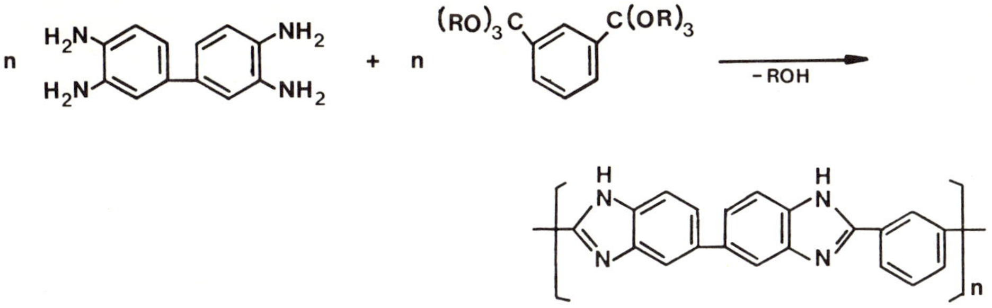

Scheme 8. Synthesis of polybenzimidazoles from bis(*o*-diamine)s and bis(ortho-ester)s

right through to the aromatized stage and does not provide for the isolation of processible prepolymers, the unusually mild polymerization conditions suggest promising development possibilities, notably for reactant systems sensitive to high reaction temperatures.

An entirely novel approach toward benzimidazole polymers was first proposed by Gray et al.[45]. Allowing 3,3′-diaminobenzidine to undergo melt condensation with 4-acetylacetophenone at 250 °C, these workers obtained a well-defined soluble polymeric intermediate through elimination of water. This prepolymer, assigned a polybenzimidazoline structure, was readily aromatized to the polybenzimidazole with elimination of methane by solid-state heating at 300 °C (Scheme 9). That the aromatization step was indeed brought about by elimination of methane rather than of the *p*-phenylene unit was demonstrated[45] by mass spectrometric monitoring of the reaction and was further confirmed by the results of a non-polymeric model reaction involving the condensation of *o*-phenylenediamine with acetophenone. This reaction (Scheme 10) produced 2-phenylbenzimidazole (by expulsion of methane) but not 2-methylbenzimidazole (through elimination of benzene).

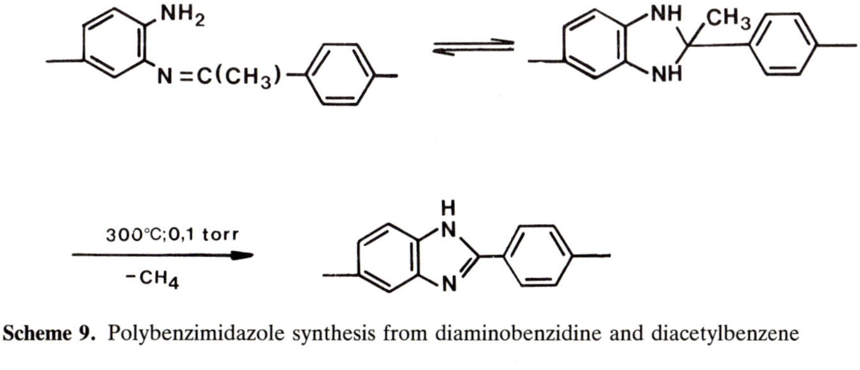

Scheme 9. Polybenzimidazole synthesis from diaminobenzidine and diacetylbenzene

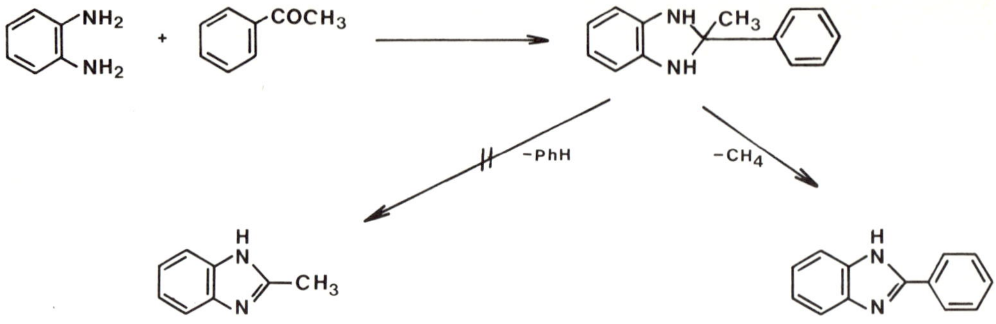

Scheme 10. Model reaction involving condensation of *o*-phenylenediamine with acetophenone

The concept of amine-carbonyl interaction as a polymerization principle was carried further by Higgins and Marvel[46]. Conducting polycondensations of bis(*o*-diamine)s with the bisulfite adduct of isophthalaldehyde in polar organic media such as dimethyl sulfoxide and amide solvents at the boiling temperatures under nitrogen (or first under nitrogen, then in air) as shown in the exemplifying sequence of Scheme 11, these investigators obtained formic acid-soluble products structurally identical with the respective polybenzimidazoles prepared by different methods. No attempts were made in this study to intercept prepolymers. Use of the bisulfite was found by the authors to be a prerequisite for successful polymerization; employment of the free dialdehyde or its acetal under the same experimental conditions favored side reactions believed to involve generation of aldehydine (Scheme 12) and other competing processes preventing clean propagation per Scheme 11. A subsequently issued, most proficiently elaborated patent by D'Alelio[47] discloses the preparation of linear benzimidazole polymers by solution polymerization, in dimethylacetamide or similar solvents, of aromatic bis(*o*-diamine)s with free dialdehydes at preferably 100–125 °C in the presence of air. According to the claim, the reaction, exemplified in Scheme 13, leads directly to the ultimate polybenzimidazole under properly controlled conditions, as the rapid oxidative cyclodehydrogenation entails ring closure before a second substitution step through Schiff base formation at the free *o*-amino group and subsequent aldehydine generation can occur.

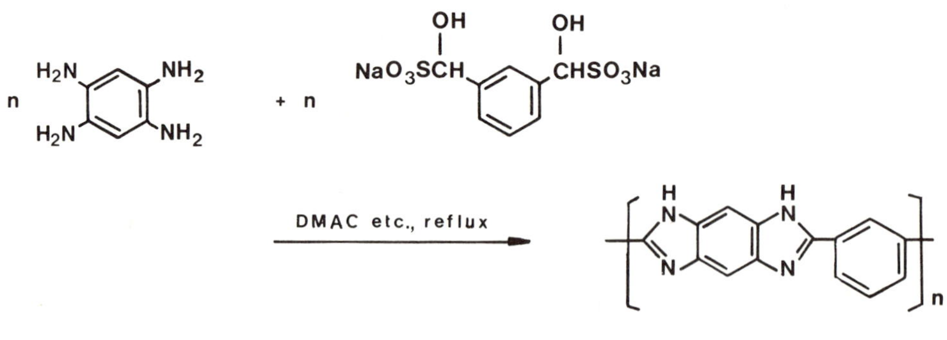

Scheme 11. Polybenzimidazole synthesis from tetraaminobenzene and bisulfite adduct of isophthalaldehyde

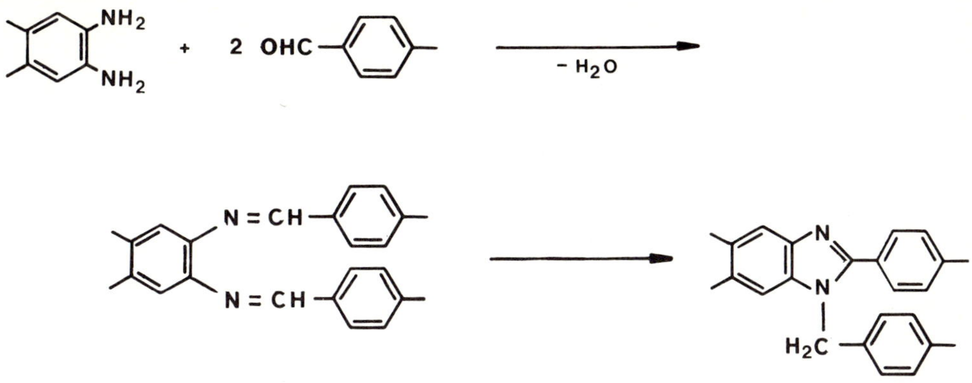

Scheme 12. Side reaction involving aldehydine formation, leading to branching and crosslinking

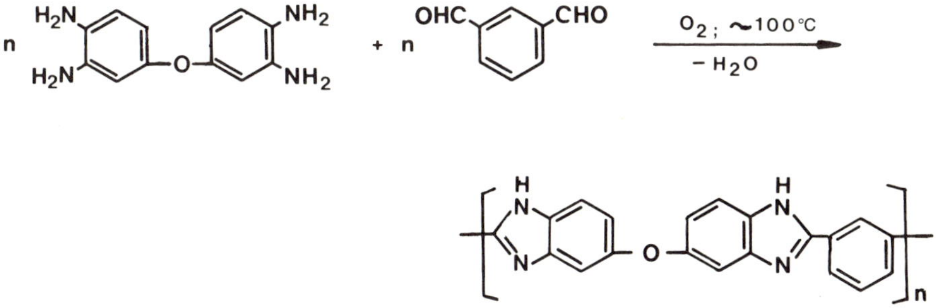

Scheme 13. Polybenzimidazole synthesis, exemplified by the direct polycondensation of 3,3′,4,4′-tetraaminodiphenyl ether with isophthalaldehyde in solution

As D'Alelio's patent embodies conditions of low condensation temperatures and so, just like some of the aforementioned solution polymerizations[44,46], teaches conditions beneficial to reactant and product integrity, it permits the preparation of polybenzimidazoles from monomers too unstable thermally or chemically for use in the melt or poly(phosphoric acid) solution processes, examples of such monomers being tetrahaloterephthalaldehydes or the crosslinkable 3-vinylisophthalaldehyde. It does not, however, provide for the isolation and application of prepolymers and, thus, for utilization of the two-stage approach so beneficial from a processing and application standpoint.

Several years ago, encouraged by the proven intermediacy of polyimidazolines in Gray's imidazole polymer synthesis[45], we set out to demonstrate experimentally an analogous intermediary stage in the polycondensation of Scheme 13 and develop a preparative approach that would enable us to isolate soluble polyimidazoline intermediates. Clearly, if such polymeric imidazolines could be isolated and their further smooth conversion to polybenzimidazoles be accomplished, the door would be opened for the development of two-stage processing techniques involving prepolymer-type materials. Initial studies in this laboratory[48], later on complemented in more elaborate investigations[49], have indeed shown that tractable prepolymers can be intercepted in the reaction of bis(*o*-diamine)s with dialdehydes, provided that the primary solution condensation is conducted at sub-zero temperatures and in the strict absence of oxygen, the solvents being of the dipolar aprotic type. Furthermore, care must be taken to ensure a stoichiometric excess of amine over aldehyde reactants during the major part of the mixing phase so as to suppress aldehydine formation (*cf.* Scheme 12); this is best achieved by adding the dilute (typically 0.1 M), deoxygenated dialdehyde solution very slowly to a rapidly stirred and deoxygenated solution of the bis(*o*-diamine) at −15 to −18 °C. After a further 15–20 h of reaction time at 0–25 °C, the prepolymer formed can be precipitated from solution and may subsequently be advanced to the benzimidazole structure by aromatization in solution under appropriate conditions. Schemes 14 and 15 illustrate this two-stage polymerization process for reactions involving 3,3′-diaminobenzidine and 1,2,4,5-tetraaminobenzene as the bis(*o*-diamine) monomers.

Spectroscopic information, supported by the results of a non-polymeric model reaction study[50], indicates the prepolymers to possess a predominantly open-chain structure of

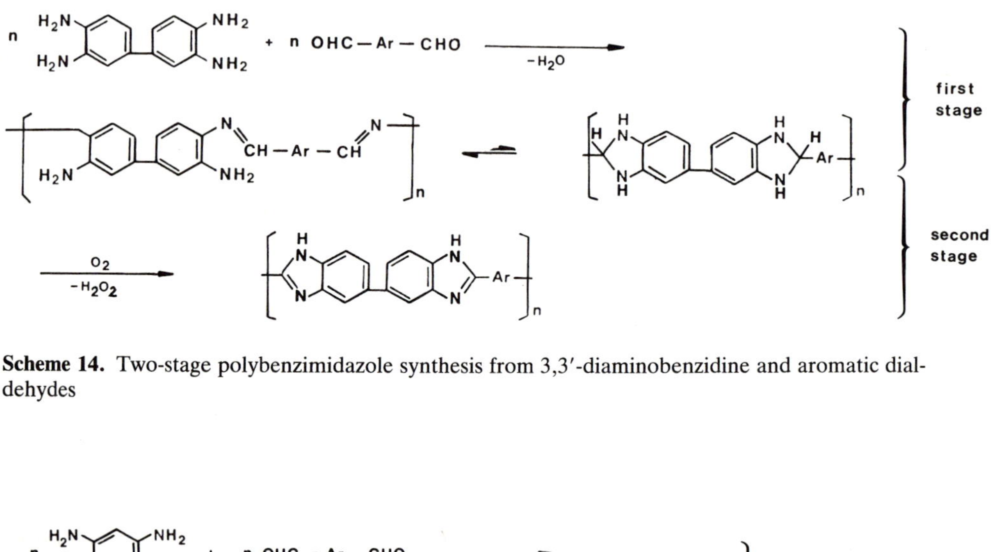

Scheme 14. Two-stage polybenzimidazole synthesis from 3,3′-diaminobenzidine and aromatic dialdehydes

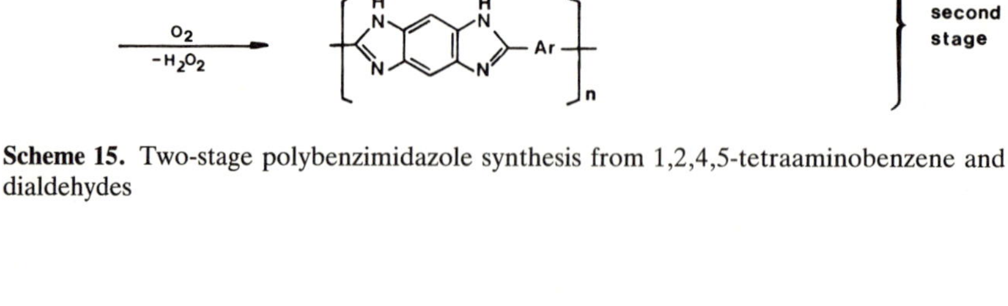

Scheme 15. Two-stage polybenzimidazole synthesis from 1,2,4,5-tetraaminobenzene and aromatic dialdehydes

the Schiff base type rather than the tautomeric imidazoline structure expected on the basis of Gray's results[45], although the subsequent aromatization step, involving a cyclodehydrogenation mechanism, most probably proceeds *via* the imidazoline tautomer continuously regenerated from the open-chain polyazomethine. In a strictly anaerobic environment, this cyclodehydrogenation is a highly inefficient process even at temperatures exceeding 150–200 °C. On the other hand, when oxidatively assisted, preferably by air, and catalyzed by certain transition metal compounds, aromatization is brought about under exceedingly mild conditions, the species eliminated probably being hydrogen peroxide[51] (rather than water). For example, in a laboratory scale, prepolymers dissolved in *N,N*-dimethylacetamide (0.2 M) containing $FeCl_3$ (2×10^{-3} M) are completely cyclodehydrogenated in 5–8 h at 60 °C by an air stream introduced at a rate of 10 ℓh^{-1}. The advantage of this two-stage synthesis, therefore, is twofold: firstly, tractable and processible prepolymers are provided, and, secondly, monomers can be employed that are too sensitive to survive the thermally or chemically demanding environments of the conventional melt or poly(phosphoric acid) solution condensation processes. Representative polymer structures falling into this latter category include the phenylenedivinylene – bridged type *6,* the metallocene-bridged types *7* and *8,* the tricarbonylchromium-complexed *10,* or the heavily chlorinated products *11* and *12* (Scheme 4), about which more will be said further below.

The progress of cyclodehydrogenation may conveniently be followed by electronic absorption spectroscopy in the low-frequency region[52]. Aromatic bis(Schiff base)s give a characteristic $\pi \rightarrow \pi^*$ band at 400–450 nm, whereas the corresponding benzimidazoles absorb strongly in the vicinity of 330–350 nm. The spectral changes associated with the conversion of an *o*-amino-substituted bis(Schiff base) to the respective bis(imidazole) are illustrated in Fig. 1 for the cyclodehydrogenation of *N,N'*-bis(2-aminophenyl)-*p*-

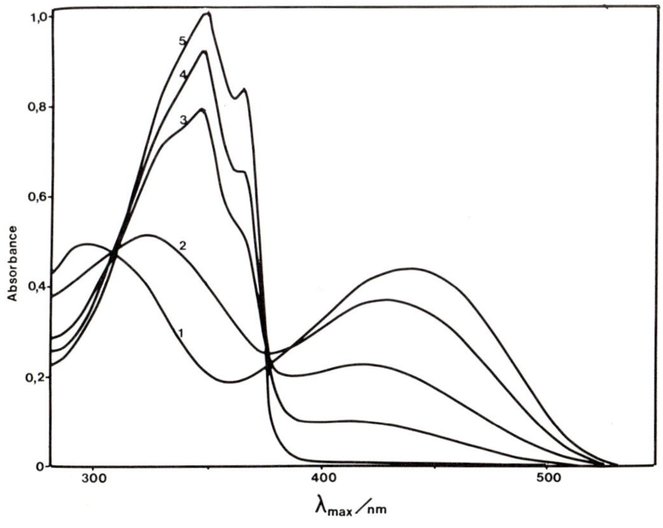

Fig. 1. Formation of 1,4-bis(benzimidazole-2-yl)benzene by $FeCl_3$-catalyzed oxidative cyclodehydrogenation of *N,N'*-bis(2-aminophenyl)-*p*-xylenediimine in dimethylacetamide at 60 °C, monitored by electronic absorption spectroscopy. Spectra were taken at time intervals over a period of 150 min. Curve 1(5): spectrum taken at start (on completion) of experiment

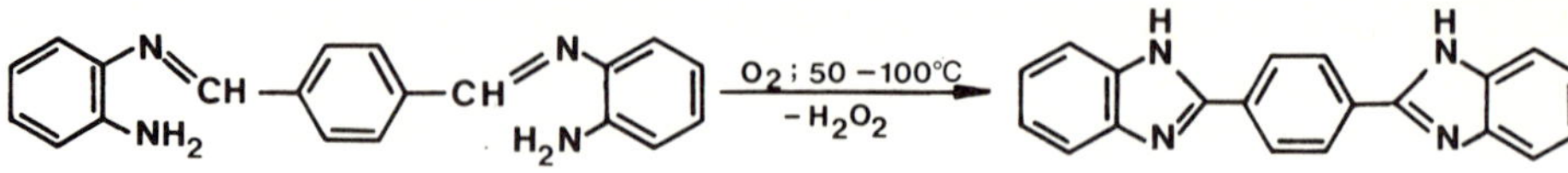

Scheme 16. Cyclodehydrogenation of *N,N'*-bis(2-aminophenyl)-*p*-xylenediimine

xylenediimine to 1,4-bis(benzimidazol-2-yl)benzene[50] (Scheme 16). The corresponding monitoring of absorbance changes in polycyclodehydrogenations leads to analogous sets of traces, except that the pertinent maxima generally appear red-shifted by some 50 nm (see 2.1); Fig. 2 depicts a representative set for the formation of polybenzimidazole *18* from the respective precursor poly(Schiff base). A more detailed examination of absorbance changes as a function of reaction time in polycyclodehydrogenations reveals high initial aromatization rates, yet extremely slow conversion towards the end of the process; indeed, the extent of imidazole ring formation amounts to some 90% at "half-time", i.e., after one-half the number of hours needed for complete disappearance of the azomethine band, and the last few percent of open-chain segments will require an unexpectedly large time fraction for aromatization. These observations suggest that, commensurate with progressively enhanced molecular chain stiffness and diminished rotational freedom,

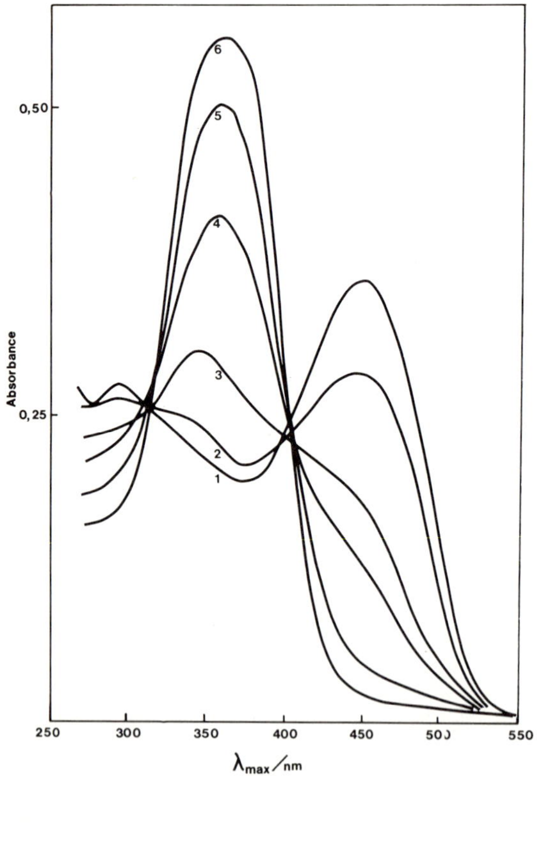

Fig. 2. Formation of polybenzimidazole *18* by $FeCl_3$-catalyzed oxidative cyclodehydrogenation of polyazomethine-type precursor polymer in dimethylacetamide at 60 °C, monitored by electronic absorption spectroscopy. Spectra were taken at time intervals over a period of 8 h. Curve 1(6): spectrum taken at start (on completion) of experiment

there is an increase in the activation energy of cyclodehydrogenation towards the end of the imidazolation process in the polymer. It is significant to note that these 5–10% of open-chain units so persistently avoiding aromatization, while observable by electronic absorption spectroscopy, cannot be detected by IR or NMR spectroscopy. As the success of monitoring absorbances by UV/VIS spectroscopy significantly depends on the existence of precursor absorption in the 400 nm region (which is generally devoid of other aromatic bands), as well as on the complete or almost complete solubility of both the ultimate polybenzimidazole and the precursor polymer in the neutral solvent system chosen for the reaction, it is not surprising to find that this technique, while of unique value in the present case, has not generally been put to use in polybenzimidazole synthesis.

Some further comments are due at this point on the problem of polymerization of highly chlorinated monomers. The synthesis of *11,* which contains a tetrachlorophenylene bridging unit, had earlier been attempted by melt polycondensation of 3,3'-diaminobenzidine with diphenyl tetrachloroterephthalate[41]; however, in contrast to the apparently smooth formation of an analogous polybenzimidazole containing only two Cl substituents on the phenylene group[16], it had not been possible to introduce more than 20% of the theoretical Cl content because of appreciable degradation at the high reaction temperatures. Even with recourse taken to the approach of Schemes 14 and 15 (Ar = 2,3,5,6-tetrachloro-1,4-phenylene)[49], certain temperature limitations must strictly be observed if clean and nondegradative polymerization is to be achieved. The precursor polyazomethines, when heated anaerobically in dimethylacetamide solution or other media for prolonged periods of time at temperatures in the neighborhood of 100 °C, or else when prepared from the monomers at this temperature level, suffer severe degradation by chain cleavage (*e.g.,* Scheme 17). Identification of 1,2,4,5-tetrachlorobenzene as one of the ultimate fragmentation products permits the inference that chain scission involves the bond connecting the aldehydic carbon atom with the phenylene group. Nonpolymeric model reaction studies performed in the author's laboratory, corroborating this view, have shown that the aldehydic carbon atom becomes incorporated into the departing hetero ring system, while a hydrogen atom is transferred from the heterocycle to carbon atom 1 of the phenylene group. This process leads to aromatization (i.e., imidazolation) of the heterocycle without concomitant dehydrogenation, and it is in fact the gain in delocalization energy on imidazolation, paired with release of the excessive

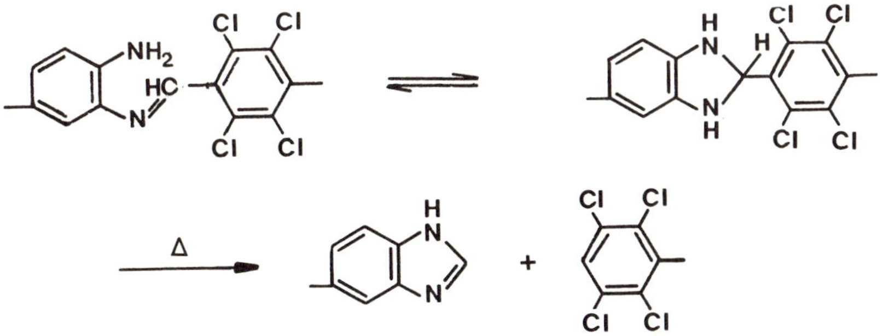

Scheme 17. Thermally induced chain scission in polyazomethine containing the tetrachlorophenylene group in the backbone

steric strain exerted by the four bulky chlorine substituents on the adjacent phenylene unit, which brings about this unusual C–C bond cleavage at the mildly elevated temperatures. On the other hand, polyazomethine formation (first stage, Schemes 14 and 15) does proceed smoothly, and for practical purposes unaccompanied by chain fission, if the polycondensation is performed at temperatures below 25–30 °C[49]. Similar temperature restrictions should be maintained during performance of the second stage. However, once converted to the imidazole structure, the polymeric chain no longer possesses the high susceptibility to cleavage as shown by the prepolymer and can moderately be heated without detriment, the reason for the enhanced thermal stability being that the driving force of aromatization no longer exists and the relief of strain along is insufficient to overcome the energy of the C–C bond connecting the units.

A discussion of synthetic approaches in the polybenzimidazole field would be incomplete without citing procedures that utilize monomers in which the imidazole ring has already been closed. Monomers of this type, hence, must possess other functional groups bringing about chain growth, whereas the imidazole ring structure is merely introduced as a preformed backbone constituent. For example, a number of benzimidazole-containing diamines has been prepared for subsequent polycondensation either with di(chlorocarbonyl) compounds, to give polyamides[53], or with pyromellitic dianhydride, giving polyimides[54], the last-named reaction proceeding *via* a well defined prepolymer stage. Similarly, benzimidazole-containing dicarboxylic acids have been used[55] for further polymerization by conventional methods. Of particular interest is the recently described[56] polycondensation of 2,6-dimethylimidazo[4,5-*f*]benzimidazole and its *N*-substituted derivatives with dialdehydes in benzoic anhydride. The reaction, which represents a Knoevenagel condensation and proceeds readily at 160–170 °C, affords polybenzimidazoles with all-*trans* vinylene groups in the chain; it does not, however, lend itself to the isolation of tractable prepolymers. An exemplifying polycondensation is shown in Scheme 18.

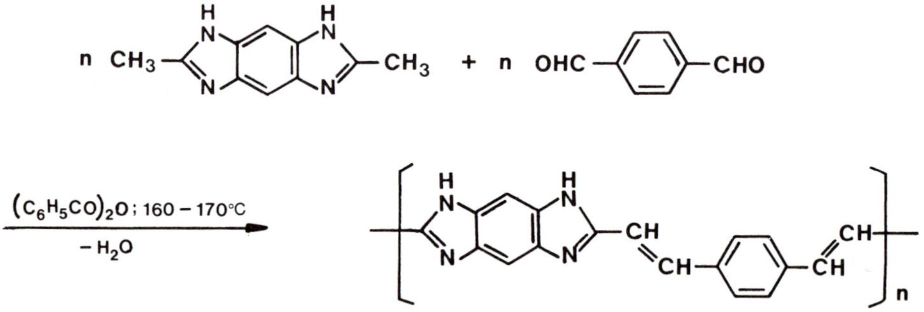

Scheme 18. Polybenzimidazole synthesis, exemplified by the polycondensation of 2,6-dimethylimidazo[4,5-*f*]benzimidazole with terephthalaldehyde

2 Properties and Applications of Polymers

In the present section typical properties of the all-aromatic benzimidazole polymers will be surveyed, and applications, as far as such have materialized, will be discussed. With thermal stability the over-riding factor in the great majority of cases where performance is assessed, one hardly finds a paper in the literature in which attention is not paid to this distinguishing feature. It has been attempted in the following to give due recognition to the emphasis so widely placed on polymer performance at elevated temperatures; accordingly, a subsection has been set aside for specific coverage of the heat resistance problem.

2.1 Chemical and Physical Properties

The all-aromatic benzimidazole polymers obtained in the synthetic efforts discussed in the foregoing section are remarkably similar in their general chemical and physical behavior, although significant deviations from the general line of characteristics (invariably towards a poorer performance level) have been observed in several instances where structural peculiarities or synthetic problems caused unsatisfactory propagation and defects in the polymer chain build-up. The products generally are yellow to brown colored solids infusible up to 300–400 °C, with some degree of crystallinity detected in the more symmetrical types[9, 30]. Although somewhat hygroscopic and tenaciously moisture-retaining even at temperatures exceeding 100 °C[41, 43, 49, 57], the benzimidazole system in the polymer chain is not prone to hydrolysis[9, 10, 26, 58]; tests conducted on the *m*-phenylene-linked bibenzimidazole polymer *1* showed this type to be unchanged after a 10-h treatment in 70% aqueous sulfuric acid or 25% aqueous potassium hydroxide solution at the respective reflux temperatures[9].

Reports on solubility behavior are somewhat controversial. Although there is consensus in the observation that most linear polybenzimidazole types, regardless of structure and origin, dissolve partly or entirely in strong protonic acids, e.g., conc. sulfuric acid or methanesulfonic aicd, contradictory observations are on record regarding the solubility properties in such weaker acids as formic acid, and in non-acidic media, such as the aprotic amide-type solvents and dimethyl sulfoxide. While some of the divergent findings can be rationalized in terms of differences in the preparative methods and experimental conditions, notably the ultimate polymerization temperatures employed, this is not so in other cases where different behavior is shown by products prepared by substantially identical procedures. For example, a polymer of type *1* prepared in poly(phosphoric acid) was found by Iwakura et al.[30] to be partially soluble in formic acid, but completely soluble in dimethyl sulfoxide and dimethylacetamide, whereas Varma and Veena[36] reported the same polymer type to dissolve completely in formic acid, yet only partially in dimethyl sulfoxide or dimethylacetamide. Similarly, for the isomeric *2*, again prepared in poly(phosphoric acid), the former group[30] found partial solubility in both dimethyl sulfoxide and dimethylacetamide, while the latter group[36] observed insolubility in the amide, and mere swelling in the sulfoxide medium, despite a somewhat lower polymer molecular mass. Nevertheless, several qualitative trends in structure-solubility relationships do emanate from a comprehensive examination of pertinent literature information.

Thus,

(i) polymers containing a *m*-phenylene group adjacent to the benzimidazole unit, such as type *1,* tend to show better solubility than their structural counterparts, such as *2,* containing the highly ordered, close-packing, and powerfully conjugating (i.e., stiffening) *p*-phenylene link[9, 30, 49]; similarly, the highly ordered 4,4′-biphenylene-bridged polymer *3* is less soluble than its more loosely packed and disordered isomer containing the 2,2′-biphenylene bridge[9], and so is the uncomplexed *2* relative to its tricarbonylchromium-complexed counterpart *10*[49];

(ii) oxygen, sulfur or sulfone bridges between aromatic units (e.g., *4* and *17–20*) enhance chain flexibility and so generally entail improved solubility[12, 14];

(iii) diaminobenzidine and tetraaminodiphenyl ether both produce more flexible and soluble polymers than does the tetraaminobenzene monomer lacking the internal C−C or C−O single-bond "hinge"; compare, for example, the dimethyl sulfoxide-soluble *1* with the corresponding *13,* which is insoluble in that solvent[9];

(iv) *N*-methylation or *N*-phenylation improves the solubility properties; thus, both *15* and *16* dissolve more readily (in dimethyl sulfoxide and formic acid, respectively) than does the unsubstituted type *13*[9, 10, 13];

(v) heavily halogenated polymers exhibit poorer solubility than their halogen-free counterparts; whereas *14,* for example, dissolves readily in sulfuric acid[9], the chlorinated type *12* is only partially soluble in that medium[49];

(vi) certain heterocyclic groups with electron donor properties, e.g., pyridine and phenoxathiin, give enhanced solubility in aprotic solvents[9, 24, 32], as does the presence of silicon[59], phthalide[60], or metallocene[49] groups in the chain;

(vii) the addition of a few percent lithium chloride to aprotic solvents tends to increase their solubilizing power[30], as does the combined application of heat (> 200 °C) and pressure[19], the last-named feature being of particular importance in fiber spinning;

(viii) polybenzimidazoles prepared under particularly mild conditions (e.g., by low-temperature solution polymerization of bis(*o*-diamine)s and dialdehydes[49], while initially distinguished by a better solubility behavior than shown by melt condensation products, tend to "age" on storage, a gradual decrease in the rate and/or extent of dissolution being noticeable; it appears reasonable to link such behavior with a loose packing arrangement of the chain segments (perhaps sustained by trapped solvent traces) at the point of polymer precipitation, this assembly slowly consolidating to a more compact and solvent-resistant state; and

(ix) published data give no indication of a viscosity (and, hence, molecular-mass) dependence of the solubility behavior of polymeric (in contrast to oligomeric) compounds.

The molecular size of polybenzimidazoles has in the pertinent literature been expressed in terms of inherent (η_{inh}) or intrinsic ($[\eta]$) viscosities, determined on sulfuric acid solutions or, less frequently, on solutions in formic acid or aprotic solvents. The effect of structure on viscosity behavior appears to be less pronounced than that of the polymerization methods used and of the monomer sensitivity to the employed reaction conditions. In general, melt polymerizations by Marvel's method give products with higher molecular mass than obtained in solution condensations, which may partly be due to increased end group reactivity and interaction at the much higher reaction temperatures encountered in the former process (Cf. Table 1). Furthermore, monomers like bis(phenoxycarbonyl)ferrocene, diphenyl tetrafluoroterephthalate, or 1,7-bis(phenoxycarbonyl)car-

Table 1. Viscosity data[a] for representative samples of polybenzimidazoles *1* and *2* prepared by different methods

Method	η_{inh}/(dl g^{-1}) *1*	*2*	Ref.
Melt and solid state[b]	3.3	1.0	9
(up to 400 °C)	≈1.0	–	19
Polyphosphoric acid soln.[c]	1.4	0.5	30
(up to 200 °C)	1.7	1.2	
	0.4	1.3	36
Sulfolane or sulfone soln.[d]	0.7	1.0	43
(up to 270 °C)			
Dimethyl sulfoxide/pyridine soln.[e]	0.8	0.8	44
(up to 100 °C)			
N,N-Dimethylacetamide soln.[f]	–	0.7	47
(up to 100 °C)	1.4	1.2	49

[a] Inherent viscosity, at 25–30 °C in 97–98% H_2SO_4 (intrinsic viscosity in Ref. 36)
[b] From 3,3′-diaminobenzidine(DAB) and diphenyl iso- or terephthalate
[c] From DAB and iso- or terephthalic acid (first line); fromDAB and iso- or terephthalamide (second line)
[d] From DAB and diphenyl iso- or terephthalate
[e] From DAB and hexaalkyl ortho-iso- or ortho-terephthalate
[f] From DAB and iso- or terephthalaldehyde in one-stage (firstline) or two-stage (second line) operation

borane, which are prone to decomposition at the high temperatures required in the melt process and so tend to upset the stoichiometry and reduce the "cleanliness" of the system, have been found to give rise to low-viscosity products[11, 17, 41)]. "Backbiting" of terminal functional groups in intermediary stages of polymerization represents another side reaction potentially militating against propagation to high molecular mass. Such reaction course can typically be expected for heteroannularly difunctionalized metallocene monomers. For example, the ferrocene-containing *7,* when obtained by low-temperature solution polymerization of diaminobenzidine with 1,1′-diformylferrocene and oxidative cyclodehydrogenation (Scheme 14), gives a viscosity value ($\eta_{inh} \simeq$ 0.5 dl g^{-1}) which, although higher than that of the melt condensation product, is lower than found for most of the non-metallocene polymers prepared under comparable conditions[49)]. A similar situation holds for the ruthenium analog *8.* Premature termination of the growing chain with formation of a macrocyclic bisazomethine end group as exemplified in Scheme 19 is a strong possibility in these cases, which finds corroboration in the results of related studies involving the reaction of 1,1′-diformyl- or 1,1′-diketoferrocenes with *o*-phenylenediamine[61)]. The exceedingly low viscosity values determined for the cobaltcenium-containing *9*[34)] may well be related to similar backbiting effects.

The viscosities obtained on a given polybenzimidazole sample may show considerable variation when measured in different solvents. Formic acid, for example, has been reported to give appreciably higher η_{inh} values than sulfuric acid[8)], although this could not be confirmed in the author's laboratory. Owing to a polyelectrolyte effect, there is some viscosity dependence on acid concentration in sulfuric acid solutions[57)]; a similar trend

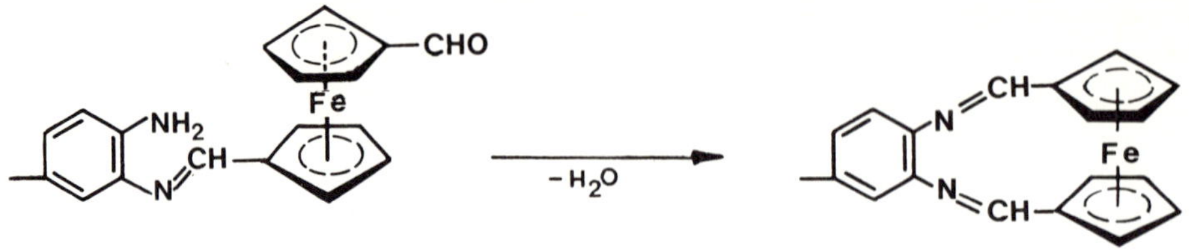

Scheme 19. Suspected chain termination in the polycondensation of diaminobenzidine with 1,1′-diformylferrocene

has been observed, and corroborated by Fuoss-Strauss plots, with formic acid as the solvent[39, 62]. Data from Marvel's laboratory[9] indicate viscosity values to be two- to three-times as high in formic acid as in dimethyl sulfoxide. Viscosities measured in the last-named medium, which is known to be strongly hygroscopic, have been found to be drastically dependent on its water content[8]. While phenomena of this kind render meaningful molecular size comparisons between products of different structural composition (and stemming from different laboratories) rather difficult, the reported results by and large agree on η_{inh} values in the approximate range of 0.4–3 dl g^{-1} (in one exceptional publication[25a] ranging up to 8–10 dl g^{-1}!). A typical polymer viscosity of 0.8 dl g^{-1} (0.5% in dimethyl sulfoxide) has been reported to correspond to a mass-average molecular mass of 54000 (determined by light scattering)[9]. This finding permits the conclusion that most aromatic polybenzimidazoles reported in the literature are characterized by $\overline{M}_w$ values in the much valued, high range of 30000–100000. In a more elaborate light-scattering and viscosimetric study, using high-molecular-mass samples of polymer *4* (prepared in poly(phosphoric acid)), Kojima et al.[63] determined from Zimm plots the mean-square radius of gyration $\langle s^2 \rangle_{LS}$, the second virial coefficient A_2, and the mass-average molecular mass $\overline{M}_w$. Representative data for runs conducted at $T = 303$ K are collected, together with respective $[\eta]$ values, in Table 2. With the aid of samples possessing $[\eta] > 1$, Kojima[62] also determined the constant $a = 1.58$ in the relationship $[\eta] = K\overline{M}_w^a$ for formic acid solutions of the same polymer *4*. The large value obtained, indicating a rod-like behavior of the polybenzimidazole chain, attests to the high intrinsic segmental stiffness, expected on structural grounds, and to an appreciable extent of uncoiling brought about by protonation in the acidic medium and concomitant establishment of imidazolium cations in the chain.

Table 2. Viscometric and light-scattering data[a] for polybenzimidazole *4*

Sample No.	$[\eta]/(\text{dl g}^{-1})$	$10^{-4} \cdot \overline{M}_w$	$10^3 A_2$	$10^8 (\langle s^2 \rangle_{LS}/\overline{M}_w)^{1/2}$/cm
1	3.2	23.3	1.25	0.835
2	2.5	15.0	1.25	0.832
3	2.0	9.7	1.56	0.829
4	1.1	4.5	1.90	0.827
5	0.9	3.6	1.82	–
6	0.7	2.9	1.32	–

[a] At 303 K, in *N,N*-dimethylacetamide. From Ref. 63. See text for symbols

The spectroscopic data of interest in polybenzimidazole characterization include IR, NMR, and electronic absorption behavior. While a comprehensive tabulation of IR frequencies for the *m*-phenylene-linked benzimidazole polymer *1* is available[8], most of the bands listed are unsuitable for identification purposes because of extensive variation from type to type. Diagnostically useful absorptions, generally of high intensity, are found for the majority of polybenzimidazoles near 800 cm^{-1} (CH out-of-plane bending, frequently overlapped by benzene-aromatic bands in the region), near 1300 cm^{-1} (C–N stretching), and in the vicinity of 1450 (in-plane vibration) and 1610 cm^{-1} (combined C=C, C=N ring vibrations). The complete IR spectra of several types have been reproduced in the literature[9, 10, 25, 30, 33, 36, 53]. Representative spectra, including that of the "work horse" polymer type *1,* are reproduced in Figs. 3 and 4.

In comparison to the plethora of published IR data, the literature information on NMR data for polybenzimidazoles is by no means abundant, partly for lack of solubilty of many polymer samples in neutral solvents, and partly also because of a tendency for line broadening in solution spectra. In an aprotic medium, typically dimethylacetamide, the common structures derived from diaminobenzidine and bridged by phenylene-containing units, such as *1–4,* give proton spectra displaying a strongly broadened imino proton signal near $\delta = 13.3$, a broad multiplet due to the aromatic protons of the tetraamine ranging from about $\delta = 7.2$ to 8.1, and, generally superimposing in part on the latter, a similarly broad multiple-band resonance region at $\delta = 7.2$–8.5 due to protons of the arylene system[49, 64, 65]. In cases such as *2,* where only a single, unsubstituted *p*-phenylene group constitutes the bridging segment, a simplified pattern results, as now the bridging-

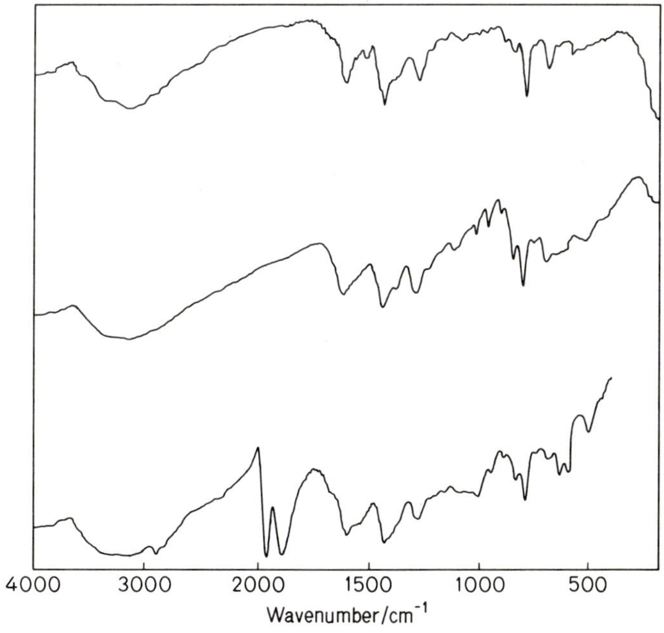

Fig. 3. Infrared spectra (KBr pellet) of selected polybenzimidazoles. From Ref. 49. From top to bottom: *1, 2, 10*

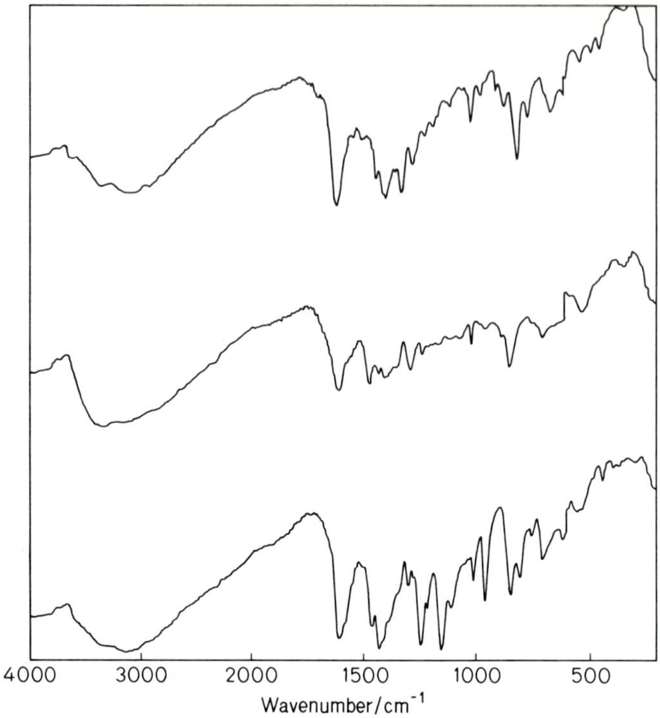

Fig. 4. Infrared spectra (KBr pellet) of selected polybenzimidazoles (continued). From Ref. 49. From top to bottom: *11, 14, 18*

group protons give rise to a single, four-proton signal emerging at $\delta = 8.48$. Comparatively simple spectra also result from incorporation of a metallocenylene group in place of arylene (e.g., *7, 8*), as the less powerfully deshielded metallocene ring protons resonate at distinctly higher fields ($\delta = 4.0$–4.7).

Proton spectra obtained on [$^2H_1$] formic acid solutions differ from those discussed in the foregoing insofar as the acid deuterium exchange with the imidazole imino protons causes collapse of the signal near $\delta = 13$; instead, a sharp singlet appears at $\delta = 10.7$. A second singlet near $\delta = 8.3$ has also been attributed to the exchanging acid medium[65].

All-aromatic polybenzimidazoles give an intense absorption band typically in the 300–400 nm region of the electronic spectrum. The absorption, frequently characterized by the appearance of high-intensity submaxima (notably on the low-energy side) in addition to the principal maximum, is almost certainly due to $\pi \rightarrow \pi^*$ transitions in the conjugated domain comprising the aromatic bridging group and the benzimidazole or bibenzimidazole unit. Spectra have generally been taken in sulfuric acid, although dimethylacetamide solution spectra have also recently been obtained. Some typical absorptions, including those for the two model compounds, 2-phenylbenzimidazole and 2,2′-diphenyl-5,5′-bibenzimidazole, are listed in Table 3. Molar absorption coefficients, ε, while not reported for sulfuric acid solutions, have been tabulated for the data obtained in dimethylacetamide.

Table 3. Low-wavelength electronic absorption maxima of selected polybenzimidazoles

Model compound or polymer	λ_{max}/nm (ε/(mol l^{-1} cm^{-1}))[a] in 97–98% H_2SO_4	Ref.	in dimethyl-acetamide	Ref.
2-Phenylbenzimidazole	300	9	306 (27 000)	50
	298	44		
2,2′-Diphenyl-5,5′-bibenzimidazole	327	9	336 (35 000)	49
	333	30		
1	336	9	339 (30 000)	49
	343	30		
	336	44		
2	379	9	383 (48 000)	49
	385	30		
	360	44		
3	368	9		
4	328	44		
10			380 (46 700)	49
11			305 (29 300)	49
13	347	9		
14	385	9		
	375	44		

[a] Principal maximum only; molar absorption coefficient ε calculated per mol of repeating unit

The λ_{max} values, as can be seen from the table, do not significantly differ for the two solvent systems used. A distinct red shift of the polymer absorption maxima relative to those of the nonpolymeric model compounds is apparent, as is a rough correlation between λ_{max} and the extent of conjugation achieved along the polymeric chain. Type *2*, for example, possessing the 1,4-phenylene link, absorbs at a higher wavelength than observed for the 1,3-phenylene-bridged isomer *1* or the ether type *4*, the latter containing an insulating oxygen bridge. A similar trend is noticed on comparing *14* with *13*. The outstanding red shift of λ_{max} of the polymeric tricarbonylchromium complex *10* indicates the bridging unit's capability of transmitting a resonance effect to be nearly as large as that of the uncomplexed 1,4-phenylene group. It is also clear from the large value of the absorption coefficient that complexation has no detrimental effect on the extent of coplanarity of benzimidazole and bridging unit achievable in solution. Finally, the exceptionally low wavelength of the chlorinated type *11* is noteworthy; steric crowding in the tetrachlorophenylene segment obviously prevents any reasonable coplanar alignment of the chain constituents and may even affect the relative orientation of the two moieties in the bibenzimidazole system.

A property of increasing concern in modern-day building and high-rise construction, plant design, and interior decoration is flammability, i.e. the propensity of most non-ceramic and non-metallic materials for ignition and combustion under realistic operating conditions. The quantitative assessment of flammability is based on the limiting oxygen index (LOI) test, which measures the minimum oxygen content (by volume) in an O_2/N_2 mixture at which the sample, contained in a particular test device, can be ignited and will support candle-like burning for a minimum of three minutes. The higher the LOI, that is, the more oxygen is required for ignition, the less flammable the material is said to be. As

Table 4. Limiting oxygen index (LOI) for representative thermoplastics[a]

Polymer type	LOI
Poly(oxymethylene)	12.2
Cotton	12.8
Cellulose acetate	14.6
Polypropylene	15.3
Polyacrylonitrile	15.3
Polystyrene	15.4
Poly(ethylene terephthalate)	15.5
Nylon 6,6	15.5
Nomex	17.0
Kapton	18.5
Poly(vinyl chloride) (unfilled)	19.5
Polybenzimidazole[b]	28.5

[a] Determined for bottom ignition. From Ref. 66
[b] Type *1*

shown by the LOI values listed in Table 4, most of the plastics engineering materials of practical interest have to be rated as flammable by common standards, their LOI being lower than the O_2 content in air (21%). In contrast, a polybenzimidazole (type *1*), with LOI = 28.5, rates as an exceptionally flame resistant material[66, 67)], vastly superior to Nomex, an aromatic polyamide, and even Kapton, the commercial grade of polyimide film. Similar results have been obtained with other all-aromatic benzimidazole polymers.

A number of mechanical and electrical properties have been reported for various benzimidazole polymers. These were generally obtained on films cast from solutions, as well as on parts molded from melt-stage intermediates, and the "work horse" in such investigations has almost always been poly(5,5′-bibenzimidazole-2,2′-diyl-1,3-phenylene) (*1*). For example, films prepared from this type were found to possess a tensile strength (ultimate) at 25 °C (200 °C) of 0.7 (0.5) g/den (units of fiber technology used) at an elongation of 7 (9)%. More illustrative of the mechanical performance are the data obtained on molded samples, e.g., tensile strength values (ultimate) in the order of 122 MN m^{-2}, reaching the 158–172-MN m^{-2} level in some cases, and a compressive strength (ultimate) of 456 MN m^{-2} (at 150 °C[68)]). Other polybenzimidazole structures, including *N*-phenylated types, by and large show a similar strength behavior. Although these data *per se* do not provide comprehensive information insofar as the benzimidazole polymers, just like other polyheterocyclics, are predominantly used for engineering purposes in the form of fiber-reinforced composites rather than neat (cf. 2.3), they demonstrate nonetheless that polybenzimidazole films or moldings possess the mechanical prerequisites for suitable materials applications. More will be said about mechanical properties in connection with thermostability. The mechanical characteristics of fibers made from this polymer will be dealt with in 2.3.

Polybenzimidazole material of the common composition is a dielectric, typically possessing a relative permittivity of 3.5 and a loss tangent of 6.5×10^{-3} in the X-band frequency region (9.375 GHz). Room temperature resistivities generally are in the vicinity of $10^{14}\ \Omega \cdot cm$. A similar behavior has been reported for *N*-phenylated polybenzimidazole structures[25,26]. Even more impressive is the dielectric performance at elevated temperatures, to be discussed in 2.2. This dielectric performance is substantially retained in composites (see 2.3).

2.2 Thermal Stability

The resistance of a material toward degradation at elevated temperatures is a feature increasingly in demand in advanced materials technologies in which operational temperature regimes are steadily raised for reasons of enhanced efficiency. It is this demand for even more increased heat resistance which, as pointed out before, has provided the most powerful impetus for heteroaromatic polymer development. The outstanding retention of useful properties in a high-temperature environment shown by the polyheteroaromatics primarily derives from the following factors:

(i) Absence of weak bonds in the backbone. Since homolytic bond scission represents an important contributory factor in the thermal degradation of a compound, it is imperative that comparatively weak bonds, including single bonds between sp^3-hybridized carbon atoms as encountered in common aliphatic compounds, be avoided in the construction of a thermally stable heteroaromatic polymer. On the other hand, single C−C bonds as constituents of conjugated systems, and even oxygen or sulfur bridges directly interconnecting aromatic units,may be tolerated in certain polymer structures, as bond stabilization through partial π-orbital overlap involving the adjacent aromatic rings generally renders such single bonds more resistant to fission than would be expected for a purely aliphatic molecular environment.

(ii) High extent of conjugation and electronic delocalization along the chain. Conjugation and, better still, fusion of aromatic ring structures leads to an essentially coplanar alignment of constituent ring systems and so may entail dramatically enhanced electronic delocalization along the polymer backbone, especially if extended π-orbital overlap results from such planar alignment. The stabilization energy so gained significantly reduces the internal energy difference, ΔE, between the polymer and its thermal degradation products (Fig. 5(A)) relative to a degradation process involving a conventional, non-aromatic polymer (Fig. 5(B)), thereby markedly decreasing the exothermicity of the reaction and, thus (neglecting entropic differences between the two types of reaction), the thermodynamic driving force for thermal decomposition. The two reaction profiles also show that there is a significantly reduced kinetic driving force for the thermodegradation of the polyheteroaromatic compound relative to that of the conventional material, a nearly doubled activation energy of decomposition, E_a, typically being associated with the heteroaromatic polymer case (Fig. 5(A)) as a result of the necessary electronic relocalization before primary bond breaking can occur.

(iii) High chain stiffness. A high stiffness, i.e., low flexibility, of the polymeric chain will clearly result from extensive conjugation and fusion of aromatic ring structures as detailed under (ii) above. This, in turn, will entail high glass transition temperatures (typically 350–400 °C and higher) and, because of a concomitant reduction in the entropy of fusion, high crystalline melting points.

(iv) "Bond healing" capabilities. Following the absorption of delocalization energy, an aromatic ring assembly *H,* now in a transition state possessing localized double and single

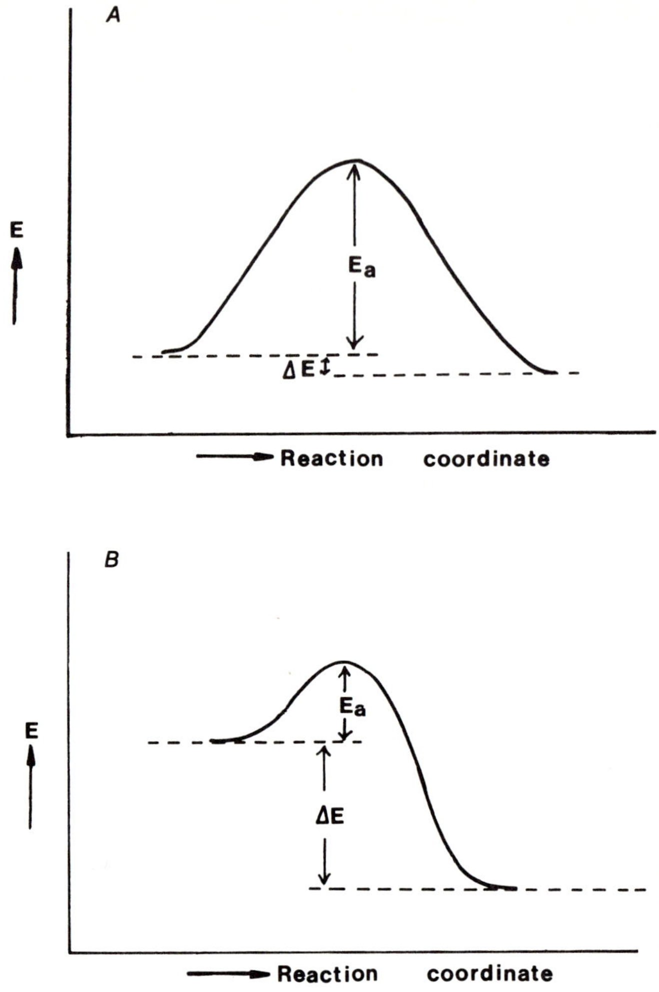

Fig. 5. Schematic internal energy vs. reaction coordinate curve for the thermal decomposition of a thermostable polyheteroaromatic (A) and a conventional polymer (B)

bonds[1] *I* may undergo cleavage at one of its single bonds, thereby converting to a biradical-type, open-chain structure *J* as shown:

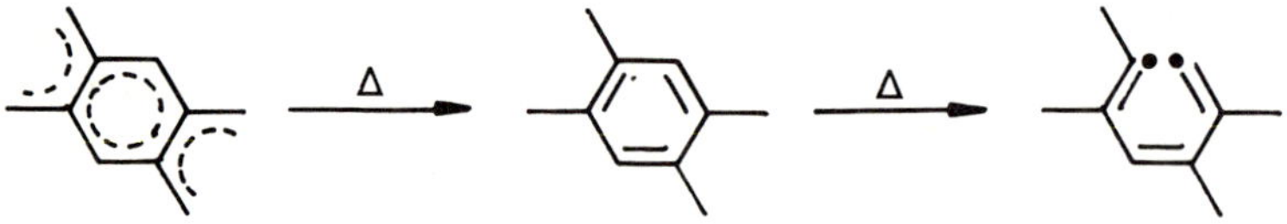

H *I* *J*

1 The structural formulae appearing in this article, although clearly implying delocalized bonds in the aromatic rings, have been drawn throughout in the classical Kékulé fashion and should not be confused with the isolated bond structure *I* given above

As the inherent chain stiffness of the system, opposing rotation out of the molecular plane, tends to retain the cisoid conformer geometry at the affected ring, the probability of bond reformation by recombination ("healing") is high, and restoration of the original ring structure may occur. For detailed treatments of the thermostability problem in polymers, the reader is referred to the more specialized literature[1, 2, 4–7, 69)].

Several analytical techniques are available for assessing a polymer's thermal stability[2, 4, 70)]. These include thermogravimetry (TG), isothermal gravimetry (ITG), and various types of thermomechanical analyses (TMA). In addition, or perhaps instead, one may simply wish to register certain mechanical properties, such as tensile or compressive strength and the respective moduli, after certain periods of specimen aging at various (and ultimately quite high) levels of temperature. In the TG test the sample as exposed to temperatures steadily increasing, typically up to 1000 °C, at a programmed heating rate, and the sample weight is continuously monitored, the relative weight loss being recorded as a function of temperature. The test, when performed under inert gas, gives information regarding the purely thermal degradation behavior at elevated temperatures and may provide valuable clues concerning mechanistic and kinetic questions. When conducted in an air environment, it provides data from which conclusions may be drawn with respect to the material's thermooxidative stability, i.e., resistance to oxygen-assisted thermodegradation. Complementing this method, ITG, in which sample weight loss is monitored as a function of time at a given elevated temperature, appears to reflect the material's thermal behavior more realistically, although, just as in TG, the assessment has its limitations, no account being taken of the actual changes in mechanical performance. The latter concern is accomodated in TMA experiments, in which a specimen's mechanical properties, for example its torsional modulus, are monitored as a function of temperature. While the results of all of these tests, taken together, will give a rather comprehensive picture of the thermal behavior, TG alone is frequently employed as a preliminary test, which has proven its value for comparative purposes, as well as for first-time evaluation studies. Accordingly, we shall find TG data reported in virtually all publications dealing with thermostability characteristics of polybenzimidazoles.

The first indication of exceptionally high heat resistance of polymers containing the benzimidazole unit emerges from the TG data presented in Marvel's pioneering paper[9)]. A TG thermogram obtained under nitrogen for the prototype *1* has been redrawn in Fig. 6, curve (a). Subsequent work in various laboratories has established a more or less reproducible TG behavior for this type, although, depending on such particular test conditions as heating rate and sample particle size, occasional deviations have been observed (cf. Fig. 6, curve (b)). The plots reflect short-time stability at temperatures as high as 550–650 °C, and even at 900 °C the residual weight of the sample (by now a heavily charred material) is quite appreciable. This performance is distinctly better than that of the highly valued polyimides and dramatically superior to that of any one of the conventional polymeric materials, which may typically show the sharp drop in the thermogram at temperatures as low as 200–250 °C and little or no residual weights in the higher temperature regime.

From Marvel's early paper[9)] comparative ITG data are also available, which show the *m*-phenylene-containing prototype *1,* after stepwise heating in nitrogen for 1 h each at 400, 450, 500, 550, and 600 °C, to have lost only a total of 4.5% of its initial weight. Very similar performance data are evident from Marvel's tabulation for the *p*-phenylene-containing polymer and various other aromatic types. The multitude of TG and ITG

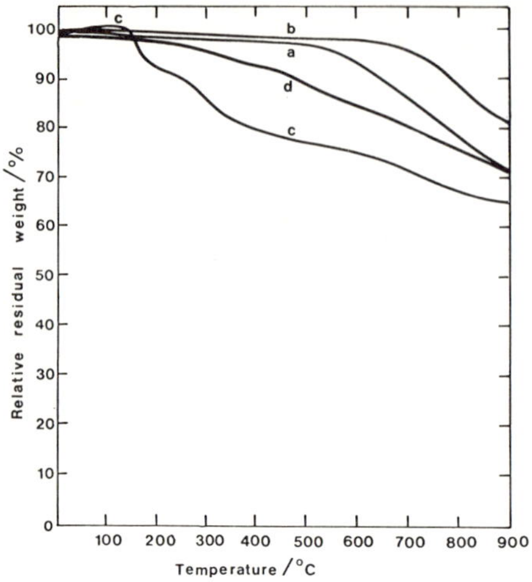

Fig. 6. TG thermograms, anaerobic. **(a)**: Structure *1;* heating rate 2.5 K min^{-1} (N_2). From Ref. 9. **(b)**: Structure *1;* heating rate 15 K min^{-1} (He). From Ref. 66. **(c)**: Structure *7;* heating rate 2.5 K min^{-1} (N_2). From Ref. 11. **(d)**: Structure *7;* heating rate 2.5 K min^{-1} (N_2). From Ref. 49

data given in the literature suggests a number of trends. Thus, differences between polymers possessing the 5,5′-bibenzimidazole ring assembly (e.g., *1* and *2*) and their analogs comprising the benzodiimidazole system (e.g., *13* and *14*) are negligible, indicating considerable conjugation across the 5,5′ C−C bond. Furthermore, no significant differences exist in the effects of *m*- and *p*-phenylene groups in the chain (e.g., *1* vs *2*), which suggests that the additional electronic delocalization realized by insertion of the more strongly conjugating *p*-unit relative to the *m*-unit is insufficient to provide a noticeable contribution to the weight loss mechanism at high temperatures. While polymers methylated at two of the four *N* atoms in the bibenzimidazole unit[13)] are practically identical in TG behavior with the non-methylated parent type, the corresponding *N*-phenyl substitution results in slightly inferior TG performance[10, 25a)], probably as a consequence of ready thermal dephenylation through fission of the comparatively weak C−N bond connecting the aryl side group to the heterocycle[25 a, b)]. Polybenzimidazoles containing the *m*-carboranylene cluster in the recurring unit show excellent thermostability[17)]; on the other hand, those containing halogenated arene units[16)] or a phosphine oxide group[37)] tend to be less stable than the prototype *1,* and so does the polymer *10* comprising the thermolabile tricarbonylchromium unit[49)]. Conflicting findings have been reported for polymers containing ether links in the chain. While Marvel's group established a clear trend of decreasing thermostability (by TG) with increasing ether link concentration in the sequence *1* (13.6% weight loss at 720 °C) → *4* (20.1% at 720 °C) → *19* (25.1% at 720 °C)[12)], Korshak's group, in a later publication[25 a)], found no indication of such a trend. Similar controversy exists for polymers containing $-SO_2-$ bridges. Another polymer type for which divergent performance data have been determined is the polymeric ferrocene complex *7.* Initial findings[11)] suggest a mediocre thermostability at best, breakdown setting in at150 °C (Fig. 6, curve (c)). It appears, however, that the inherent weakness of the backbone must be traced to structural defects in the chain, introduced as a consequence of the high condensation temperatures employed, rather than to the presence of the metallocene complex *per se.* This follows from the appreciably higher thermostability observed[49)] for a polymer of the same structure *7* prepared by the low-temperature, two-stage approach of Scheme 13 (Fig. 6, curve (d)). An even more gratifying performance is shown by the ruthenocene-containing analog *8* prepared by the same technique[49], relative residual weights exceeding those for *7* by 8–10% in the critical temperature range of 500–1000 °C. The outstanding performance of *8* is well in accord with the known heat stability of the ruthenocene complex in other polymeric compounds[71].

The mechanism of pyrolytic decomposition, i.e., thermal breakdown in the absence of oxygen, has been proposed[72] to comprise primary hydrolytic scission of the imidazole ring system by covalently held water, producing an amide linkage. This process, depicted in the first two lines of Scheme 20, occurs at temperatures below ca. 550 °C. At about 600 °C homolytic stepwise scissions take place with evolution of HCN and NH_3, leading to segments such as those shown in the fourth line of Scheme 20. At temperatures in the region of 625–800 °C, CO and H_2 evolution from intermediary amide structures produces more highly fused *N*-containing ring structures of the type shown in Scheme 21; these represent typical segments as found in the charry residues generated in polybenzimidazole pyrolysis.

As one might expect, a different picture emerges for the thermooxidative stability as reflected in the results of TG and ITG tests performed in air. The participation of oxygen in the overall degradation process leads to the intermediacy of imidazole free radicals and oxygenated free radicals with concomitant chain scission[73], and the net result is a rather catastrophic weight loss observed as the test temperature approaches the 450–500 °C level. The representative TG thermograms of Fig. 7, obtained in air, illustrate this trend

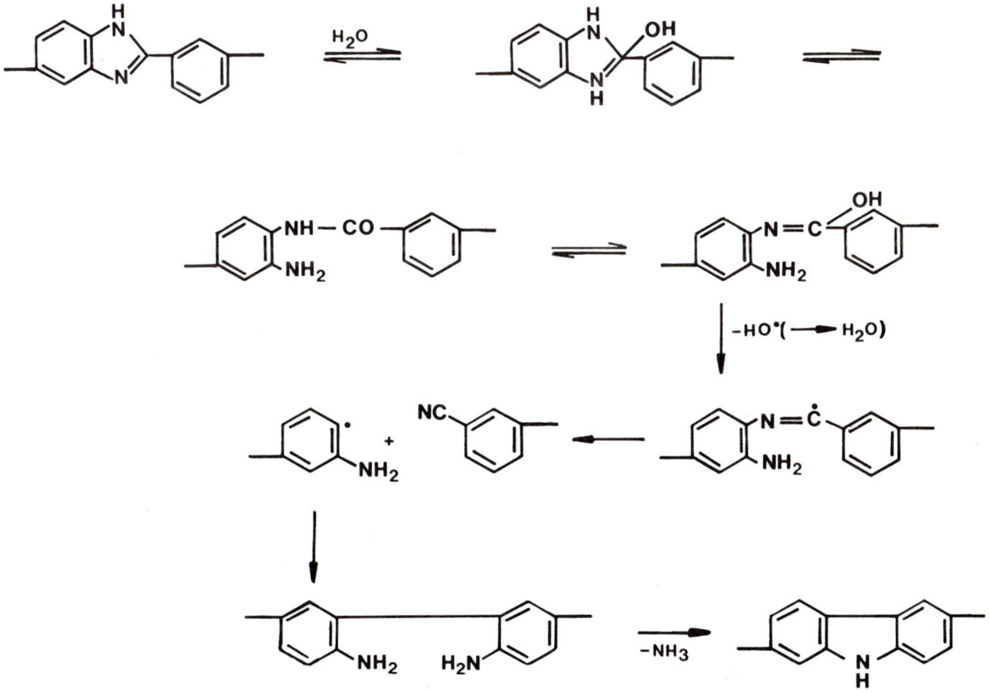

Scheme 20. Pyrolytic decomposition of a polybenzimidazole in temperature region below ≈ 625 °C

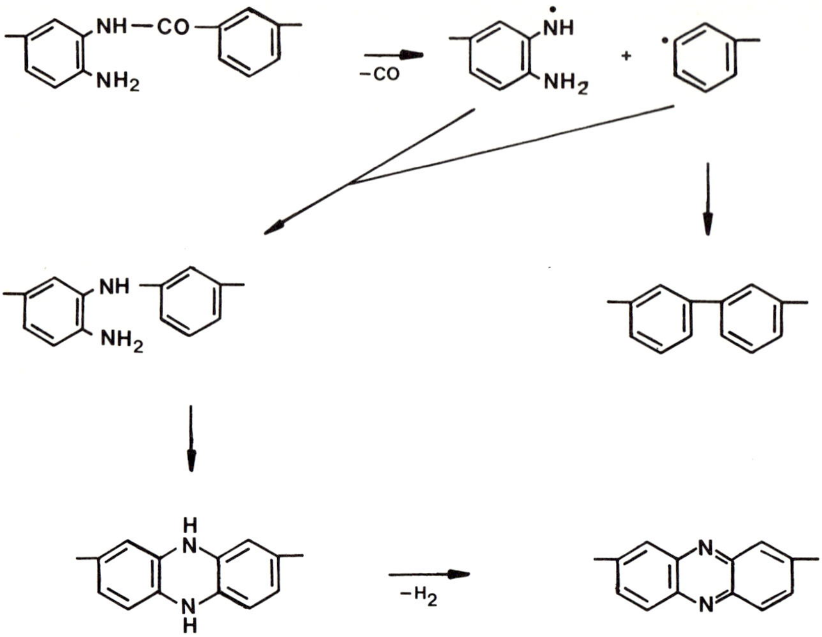

Scheme 21. Pyrolytic decomposition of a polybenzimidazole in the temperature region 625–800 °C

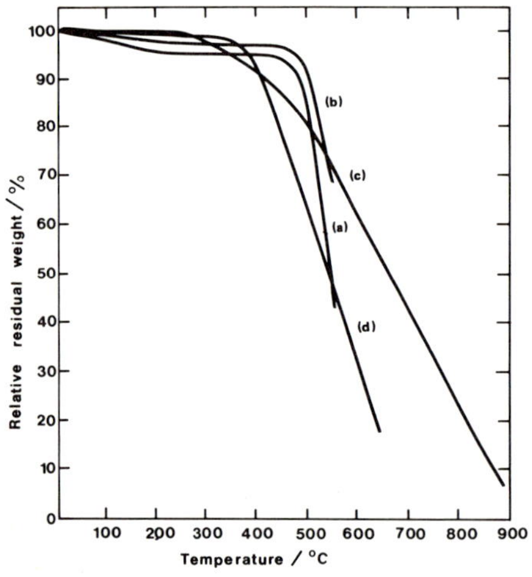

Fig. 7. TG Thermograms, in air. **(a)** Structure *1;* heating rate 3.0 K min^{-1}. From Ref. 30. **(b)**: Structure *2;* **(b)**: Structure *2;* heating rate 3.0 K min^{-1}. From Ref. 30. **(c)**: Structure *15;* heating rate 2.5 K min^{-1}. From Ref. 13. **(d)**: Structure *16;* heating rate 2.5 K min^{-1}. From Ref. 10

convincingly. More detailed comparisons show the differences in the oxidative TG test performance between *m*-phenylene- and *p*-phenylene-bridged polymers, and also between *N*-phenylated types and their unphenylated counterparts, to be negligibly small, indicating that within the comparatively short exposure times at high temperatures the effect of oxygen participation is not pronounced. In strong contrast, ITG tests conducted at 300 and 400 °C reveal a significantly higher thermo-oxidative stability of the *N*-phenyl-substituted polymers[8)]. The same trend has been reported for a polybenzimidazole similar to structure *16,* except with the central fused benzene ring replaced by a fused pyridine ring[24)]. Since ITG, as stated before, represents a more realistic assessment of the thermo-oxidative behavior under actual application conditions comprising longer exposure to high temperatures, one must conclude that, on balance, the *N*-phenylated compounds are superior to their unphenylated parent polymers in their resistance to combined thermal and oxidative attack. In light of the imidazole free radical intermediacy, brought about by initial hydrogen atom extraction from the imidazole ring of an unphenylated compound, this result is in accord with expectation. A further finding concerns the relative thermo-oxidative stability of ether and sulfone bridges in a benzimidazole-containing polymer backbone; the presence of such bridging groups, while not significantly affecting TG results, entails a distinct reduction of thermo-oxidative stability when measured in ITG[25a)].

It is instructive to compare the TG and ITG performance of polybenzimidazoles with their behavior in mechanical property testing at different temperatures. Quite excellent high-temperature performance in air is reflected in the compressive strength data collected in Table 5 for a molded material of type *1*[68)].

It is seen that catastrophic failure did not occur until the temperature exceeded ≈400 °C. Equally impressive are the strength data presented in Table 6, which were obtained on molded specimens after various periods of heat aging at 316 °C in air[68)].

Table 5. Compressive strength of molded polybenzimidazole[a] determined at different temperatures

Test temperature/°C	Compressive strength[b]/(MN m^{-2})
149	456
202	407
260	401
316	403
371	372
428	279
538	5

[a] Type *1*.
[b] Ultimate; data converted from psi units. From Ref. 68

Table 6. Compressive strength of a molded polybenzimidazole[a] determined after different heat aging periods[b]

Aging period/h	Compressive strength[c]/MN m^{-2}
1	459
200	383
300	362
500	366
800	291
1000	325

[a] Type *1*
[b] Aging in air at 316 °C
[c] Ultimate; data converted from psi units. From Ref. 68

Table 7. Relative permittivity and loss tangent[a] of molded polybenzimidazole[b] determined at different temperatures

Test temperature/°C	ε	$\tan \delta \times 10^3$
Room temperature	3.48	6.5
316	3.58	7.0
428	3.64	8.1
538	3.66	8.8
668	3.80	19.1

[a] Both determined at X-band frequency (9.375 GHz). From Ref. 74
[b] Type *1*

There is a similarly outstanding retention of dielectric properties at elevated temperatures. Typical values for the relative permittivity and loss tangent at different temperatures are listed in Table 7[74)]. Some mechanical properties of polybenzimidazole composites tested at high temperatures will be discussed in the subsequent section.

2.3 Applications

Although, for reasons of economics, as well as existing difficulties at the present state of polybenzimidazole manufacturing and processing technology, no large-scale commercialization is in sight, technical benzimidazole polymer steps and their respective prepolymers in the form of partly polycondensed mixtures have for several years been in use as engineering materials in military and aerospace hardware production; most of the fundamental development work done is documented in government technical reports, some of these with restricted access. In addition, certain types have been on the commercial market in small volume for specialty applications where their currently high manufacturing costs are accepted in view of their excellent performance. Some typical fields of application are briefly discussed in the following.

Structural engineering materials. Use as structural materials, especially under conditions of ablation or in high-temperature environments, has been one of the prime objectives of polybenzimidazole development ever since Marvel's first publication.This holds to a small extent for molded products and to a predominant extent for laminates. Almost all the polymers evaluated and used in these technical applications are of the type *1*.

It was shown in the preceding section that molded polybenzimidazole possesses outstanding mechanical and dielectric properties at elevated temperatures and/or after heat aging in air. Accordingly, polybenzimidazole moldings have found use, although as yet limited, as structural components in automative, aeronautical, and electrical engineering applications where these properties are of importance. Considerably more development emphasis has been placed on composite materials made of fiber reinforced laminates. The prepregs used for lamination are preferably manufactured by hot melt techniques, which circumvents the need for the removal of large solvent volumes as required in the conventional varnish impregnation technique[75)]. Both glass and quartz fiber materials have been employed for reinforcement. The excellent work by Reed and Hidde[74] on composite behavior in terms of flexural, tensile, and compressive strengths and moduli as a function of temperature, subsequently reviewed by Levine[8], demonstrates that up to about 300 °C there is hardly any decrease in modulus and but little or moderate decrease in mechanical strength, the least change being apparent in the tensile mode. Since the relative permittivity is little affected by temperature[74], it stands to reason that polybenzimidazole laminates, especially those containing pure silica fiber reinforcement, are excellent materials for radar-transparent radome construction, competing with the polyimides in this application. Other composite designs, exemplifying more recent developments, have been described in a technical report[76] and elsewhere[77–79]. The specialized use of polybenzimidazole composites in ablation, while unpromising in early attempts when porosity still presented a major processing problem, has since been further investigated and found to be feasible, especially in view of the very high char yields obtained[80]. The ablative performance of an *N*-phenylated polymer has also been reported[81].

Thermo-insulating materials. Polybenzimidazole material lends itself exceedingly well to heat- and sound-insulating applications in the form of foams or loosely textured mats. Prepolymer may typically[76] be molded, after compounding with carbon fibers or phenolic microballoons, at low temperatures, e.g., 120 °C; the aromatization reaction is then completed at higher temperatures, and the specimens are postheated ("cured") at about 500 °C. Foams of this type are of rather high density, typically 0.5 g cm^{-3}, but offer high compressive strength and excellent load-bearing characteristics, desirable for structural foam applications at high temperatures. A detailed survey of properties of a commercial molded foam can be found in Levine's article[8]. A more recent paper[20] features the foaming of unfilled and filled prepolymer (prepared from 3,3′-diaminobenzidine and diphenyl isophthalate), which is slowly heated in a suitable pan under nitrogen to about 500 °C. Foaming occurs mainly between ca. 185 and 275 °C. Typical foam densities attained under these conditions are in the remarkably low ranges of 0.03–0.04 g cm^{-3} (unfilled) and 0.05–0.13 g cm^{-3} (3–8% filler content), and the materials possess excellent mechanical properties at both low and high temperatures in addition to showing exceptionally low flame spread, self-ignition, and smoke generation characteristics. Foams of this low-density type are therefore excellent candidate materials for fire and thermal insulation in automotive, aircraft, and aerospace vehicles, for ablative use, as well as for combined sound and high-temperature insulation in jet engine nacelles and similar applications where both heat and high noise levels are to be coped with. Many other foam and matted polybenzimidazole materials have been disclosed in the patent literature, two exemplifying cases being cited here[82].

Electro-insulation materials. The retention of dielectric properties in a high-temperature environment, coupled with good corrosion resistance in contact with certain reactive chemicals, suggests excellent possibilities of polybenzimidazole use in electrical insulation and other dielectric applications at high operating temperatures and/or in aggressive chemical environments. Typical applications, hence, can be found in special cable and wire insulation, in the manufacture of circuit boards and radomes for supersonic aircraft, as battery and electrolytic cell separators, and as fuel cell frame structural materials. Some recent publications in the patent and technical report literature[83] may serve to illustrate such applications.

Structural adhesives. The steadily increasing use of structural adhesives in metal joining has prompted numerous development efforts, mostly in aircraft and space design-oriented organisations, aimed at the employment of heat resistant organic adhesives for stainless steel, aluminum, and titanium structural component bonding and honey-comb fabrication. A comprehensive account[84] covering the use of a commercial polybenzimidazole-type adhesive has been reviewed by Levine[8] and Hergenrother[85]. Although excellent bonding characteristics have been determined not only at high temperatures but even in the cryogenic region down to – 196 °C and lower[86], where most organic adhesives undergo strong embrittlement, commercial applications have remained few until now. The reason for this poor acceptance by industrial processors is found, firstly, in the high costs of the adhesive and, secondly, in the requirement of hot melt processing, which is difficult to control and does not lend itself well to routine and large-scale manufacturing techniques. The use of polybenzimidazole materials as adhesives is, therefore, largely restricted at present to military and aircraft component design.

Fibers and fabrics. The good spinning characteristics of polybenzimidazoles suggest applications in the fibers and fabrics field. After early studies of spinning behavior and fiber properties had demonstrated an excellent potential[57], numerous efforts have been disclosed subsequently in the journal and patent literature aiming at optimizing spinning parameters and fiber characteristics, the cited publications[66, 87] and disclosures[88] being representative. Polybenzimidazole, preferably of the type *1*, is most conveniently made into fibers by dry-spinning from dimethylacetamide solution containing a few percent lithium chloride for solubility improvement. The washed and dried yarns are generally drawn for enhanced orientation and strength, drawing temperatures for high-performance materials being as high as 400–500 °C. Drawn filaments may typically possess tenacities of 4.5–6 g/den and elongations of 19–27% (for comparison, average values for regular grade of cotton fiber are 2–3 g/den and 5–10%), and these properties do not change markedly at temperatures up to about 300 °C. In addition to being nonflammable, the fibers show an outstanding moisture regain (water content in equilibrium with surrounding air of 65% relative humidity at 21 °C) of some 13%, most favorably contrasting with acrylics (2%) or Nylons (4.5%). Furthermore, fabrics woven from polybenzimidazole yarns possess a good "hand" and are comfortable to wear. It is for these reasons that polybenzimidazole textiles are being used advantageously for protective clothing, e.g., in special fire-fighting suits and in flight crew clothing for military and space missions. Similarly effective applications are suggested for aircraft interior decorating, e.g., in seat cover design and carpeting[89]. It is of interest to add that polybenzimidazole yarns were used extensively in the American Apollo moon landing program, for example, in seat belts, portable life support systems, and in the umbilical cord with which "space walking" astronauts were tied to the space craft. For more mundane applications requiring lower materials costs, polybenzimidazole fibers may be blended with the less expensive, more conventional fiber types[90].

Interesting applications of polybenzimidazole fibrids and ultrafine fibers, developed recently under NASA contract[91], include their fabrication into ultrathin paper with a target weight of 3.5 g m^{-2} for subsequent metallization, and into mats replacing asbestos for use as separators in fuel cells. Another challenging application involves the carbonization of polybenzimidazole monofilament at 1100 °C to give a high-strength carbon fiber for use as reinforcement in advanced composites technology[92]. Continuous polybenzimidazole yarn can similarly be graphitized at 2800–3000 °C, yielding a high-strength, high-modulus graphite fiber. The process is superior to conventional cellulose or polyacrylonitrile yarn graphitization as the usual preoxidation treatment prior to pyrolysis can be appreciably shortened[93)].

Reverse osmosis membranes. The exceptionally high moisture regain observed with polybenzimidazole fibers prompted a team at Celanese Research Co to investigate the utility of polybenzimidazole films as semipermeable membranes for reverse osmosis processes, such as sea water desalination[66, 94)]. A continuous process was devised in which films were cast from solution into a water precipitation bath. The films were tested for reverse osmosis performance with a saline solution (0.5% NaCl) as feed stream at a pressure of 4.14 MN m^{-2} and a flow rate of 19.8 m min^{-1}. Salt rejection was ca. 95% throughout. A cellulose acetate film of the type commonly used as a reverse osmosis standard was tested under the same conditions for comparison. Table 8 shows the results.

It is seen that at temperatures in the 20–50 °C region both membranes showed substantially the same performance in terms of provided flux. However, the polybenzimidazole film performed increasingly better as the temperature was raised further, the flux increasing from 693 l m^{-2} day^{-1} at 49 °C to 1019 l m^{-2} at 90 °C. In contrast, the cellulose derivative standard showed a sharp drop in flux to about one-half over this temperature span and turned out to be entirely impermeable at 90 °C. A similarly outstanding performance at the even higher test pressure of 6.89 MN m^{-2} was observed with hollow filaments spun from polybenzimidazole solutions for this specific purpose in the same development program[66)]. As such hollow filaments, when bundled and co-aligned in a suitable tube module, present a higher specific operating surface and tolerate a higher pressure in relation to a membrane, the hollow-filament approach should offer a considerable economic advantage over the membrane technique. Although there is still a number of problems to be solved, such as long-term performance without flux decrease, the findings of this[66, 94)] and other studies[95)] for both membrane and hollow filament material clearly demonstrate the superiority of polybenzimidazoles in high-temperature reverse osmosis, where the more conventional polymeric membrane materials would prove a poor or totally unsuitable choice.

Table 8. Performance of cellulose acetate and polybenzimidazole membranes in reverse osmosis test[a]

Operating temperature /°C	Testing time/h	Flux/(l m^{-2} day^{-1}) Cellulose acetate	Polybenzimidazole[b]
21	5	489	571
49	5	815	693
75	20	204	774
90	5	0	1019

[a] Flux data, converted from US system of units, are those measured at end of testing time. See text for other conditions. From Ref. 66

[b] Type *1*

3 Summary and Conclusions

The classical process of polybenzimidazole synthesis from aromatic bis(*o*-diamine)s and the phenyl esters of dicarboxylic acids, designed by Marvel and collaborators some 20 years ago and until now the preferred preparative method, gives polymers of high molecular mass in unsurpassed yields. However, proceeding in the melt and subsequently in the solid state at temperatures up to 400 °C, it is restricted in its applicability to monomers and end products sufficiently stable to survive this harsh thermal environment. Iwakura's poly(phosphoric acid) solution polymerization method offers the twofold advantage of proceeding nearly quantitatively at appreciably lower temperatures (180–200 °C) than required in the melt process and utilizing the more readily available free diacids in place of the phenyl ester monomers. It tends, however, to give rise to phosphorus-containing products; in addition, no monomers sensitive to an acidic environment at these temperatues can be used. Numerous other solution polymerization techniques employing bis(*o*-diamine)s and dicarboxylic acid derivatives have since been made available in which essentially neutral solvents, preferably of the dipolar aprotic type, are used. Doubtlessly the most promising of these, and certainly the one demanding the least severe reaction conditions, is Vogl's method employing bis(ortho-ester)s as the diacid monomers; essentially complete conversion is brought about in these polycondensations within a fraction of 1 h at temperatures as low as 100 °C. The only major drawback of this procedure obviously derives from synthetic hurdles, as the preparation of the bis(ortho-ester)s is by no means simple and straightforward. Other reactions leading to polybenzimidazoles include the solution polymerization of bis(*o*-diamine)s with dialdehydes or dialdehyde derivatives designed by Marvel and independently by D'Alelio; both reactions proceed at temperatures as low as those required in the bis-(ortho-ester) polycondensation and offer the additional advantage that the dialdehyde monomers in general are more readily accessible than the corresponding bis(ortho-ester)s.

While the synthetic approaches summarized in the foregoing all proceed to the ultimate stage of aromatization and thus do not offer the benefit of prepolymer utilization in processing, several solution polymerization techniques featuring such prepolymer intermediacy are available. These include the low-temperature polymerization of bis(*o*-anilinoamine)s and bis(*o*-acetamidoamine)s with bis(acid chloride)s and subsequent thermal cyclodehydration of the poly(aminoamide) intermediates. The polybenzimidazoles formed in this type of reaction are thus partly *N*-substituted, the consequences of which have been discussed. The low-temperature solution polymerization of bis(*o*-diamine)s with dialdehydes constitutes the latest development in two-stage polybenzimidazole synthesis, allowing for the isolation of tractable polyazomethine intermediates and their further-advancement by oxidative cyclodehydrogenation to the fully aromatized benzimidazole polymers. As both steps proceed in a neutral medium at temperatures below 100 °C, monomers sensitive to heat or aggressive chemical environments may safely be employed. The molecular mass values attained tend to be somewhat lower than observed in the conventional melt and poly(phosphoric acid)solution polymerizations, where the considerably higher reaction temperatures bring about a more efficient end group interaction. In this regard, however, one may expect future improvements through optimization of first-stage polycondensation conditions. Aromatic 1,2-bis(trimethylsilylamine)s are known[96)], for example, to react smoothly and rapidly with aldehydes to give azomethines (and thence, benzimidazoles), and one should be able to utilize this reaction to advantage for high-molecular-mass polymer formation.

Turning to properties and applications now, we find the aromatic benzimidazoles to represent an outstanding class of engineering materials for use in a variety of applications over a wide temperature span. The high chemical stability and excellent heat resistance, non-flammability, good mechanical, dielectric, adhesive, and fiber properties all combine to render the polybenzimidazoles the materials of choice for structural use, lightweight thermoinsulation and advanced engineering design, as well as in dielectrics and heat resistant wire coatings, fuel cell and battery components, metal hot melt adhesives for use at both high and low temperatures, thermally stable and flame-retardant fabrics and construction assemblies, and, lastly, in high-performance membranes for reverse osmosis applications.

It should be evident from the data presented in the foregoing sections that certain selected properties can be enhanced and emphasized by suitable structural modifications. Thus, while the polybenzimidazoles unsubstituted at the hetero atoms show some sensitivity to oxygen at elevated operating temperatures relative to other polyazoles devoid of hydrogen atoms at the heterocycle, their long-term thermo-oxidative stability will be improved both by phenylation of the imino nitrogen atoms and through intermolecular crosslink formation between such imino nitrogen atoms, e.g., *via* methylene bridges generated from pre-introduced *N*-(hydroxymethyl) groups[97]. The general thermal and thermo-oxidative stability level can probably also be raised still further by a mild crosslinking involving reactive end groups (e.g., –C≡CH, –CN), although such treatment may not necessarily improve the mechanical strength, which depends *inter alia* on the existence of hydrogen bonds originating from the imino nitrogen atoms. Another strategy aimed at the further enhancement of chain rigidity and thermal or thermo-oxidative stability could involve the introduction of vinylene bridges, as in *6*, followed by anaerobic heating at temperatures exceeding 300 °C, which should result in additional aromatization through ring fusion with accompanying elimination of hydrogen. Enhanced resistance to flammability and flame propagation can be achieved by halogen substitution, as in *11* and *12*, preferably coupled with additional introduction of synergistic P or Sb atoms, although the overall heat resistance of such structures is likely to suffer. The possibility of introducing η^6-bonded transition metal atoms, as in *10*, opens up an alley toward thermally stable polymeric catalysts in view of the proven catalytic activity of nonpolymeric tricarbonylchromiumarenes[98]. Structures of type *10* may also be useful for the introduction of porosity through high-temperature expulsion of carbon monoxide, which could have a bearing on the performance of polymeric membranes. The numerous technologically useful properties of the metallocenes suggest interesting features of the ferrocene-containing *7*, notably in the areas of redox and electron transfer phenomena, catalytic activities, triplet quenching, and photosensitization.

In summary, the polybenzimidazole story reviewed in the foregoing sections reflects a vast research effort indeed, accomplished over the years in numerous polymer laboratories. A wide range of direct and two-stage synthetic approaches is now available to the polymer chemist, and an equally wide scope of challenging possibilities of application has opened up to the technologist. It should also be clear from the presented material, however, that the stage of commercial maturity has not as yet been attained, and intriguing new developments both on the preparative side and in the application sphere can, therefore, with confidence be expected in the years to come.

Acknowledgment. Much of the work performed in the author's laboratory and cited in this article was generously supported by *SASOL LTD.*, for which thanks are due to Mr. *J. A. Stegmann*, Managing Director.

References

1. Mulvaney, J. E.: Encycl. Polym. Sci. Tech. *7*, 478 (1967)
2. Frazer, A. H.: High Temperature Resistant Polymers, Wiley, New York 1968
3. Koton, M. M.: Advan. Macromol. Chem. *2*, 175 (1970)
4. Pezdirtz, G. F., Johnson, N. J.: Thermally Stable Macromolecules, in: Chemistry in Space Research, Landel, R., Rembaum, A., Eds., Elsevier, New York 1972
5. Cotter, R. J., Matzner, M.: Ring-Forming Polymerizations, Pt. B, Vol. 1, Heterocyclic Rings, Academic Press, New York 1972
6. Neuse, E. W.: Mater. Sci. Eng. *11*, 121 (1973)
7. Cassidy, P. E.: Thermally Stable Polymers, Dekker, New York 1980
8. Levine, H. H.: Encycl. Polym. Sci. Tech. *11*, 188 (1969)
9. Vogel, H., Marvel, C. S.: J. Polym. Sci. *50*, 511 (1961); Marvel, C. S., Vogel, H.: US Pat. 3,174,947 (1965); Chem. Abstr.: *63*, 7137 (1965)
10. Plummer, L., Marvel, C. S.: J. Polym. Sci., Part A *1*, 1531 (1963)
11. Plummer, L., Marvel, C. S.: J. Polym. Sci., Part A *2*, 2559 (1964)
12. Foster, R. T., Marvel, C. S.: J. Polym. Sci., Part A *3*, 417 (1965)
13. Mitsuhashi, K., Marvel, C. S.: ibid. pg. 1661
14. Lakshmi Narayan, T. V., Marvel, C. S.: J. Polym. Sci., Part A-1, *5*, 1113 (1967); Bracke, W., Marvel, C. S.: J. Polym. Sci., Part A-1, *8*, 3177 (1970)
15. Korshak, V. V. et al.: Dokl. Akad. Nauk USSR *149*, 104 (1963); Korshak, V. V. et al.: Vysokomol. Soedin., Ser. B *6*, 1251 (1964); Izyneev, A. A., Mazurevskii, V. P., Korshak, V. V.: Vysokomol. Soedin., Ser. A *17*, 1200 (1975)
16. Korshak, V. V., Izyneev, A. A., Mazurevskii, V. P.: Dokl. Akad. Nauk SSSR *220*, 372 1975)
17. Green, J.: Symp. High Polymers, Am. Chem. Soc. West. Reg. Meet., Los Angeles, Cal., Nov. 1965; Proc. L-1; Green, J., Mayes, N.: J. Macromol. Sci., Chem. *1*, 135 (1967)
18. Gitina, R. M. et al.: Vysokomol. Soedin., Ser. B *9*, 447 (1967)
19. Conciatori, A. B., Chenevey, E. C.: Macromol. Synt. *3*, 24 (1968); US Pat. 3,433,772 (1969); Chem. Abstr.: *70*, 97414 (1969); Conciatori, A. B. et al.: J. Polym. Sci., Part C *19*, 49 (1967)
20. Kourtides, D. A., Parker, J. A.: Polym. Eng. Sci. *15*, 415 (1975)
21. Breed, L. W., Wiley, J. C.: J. Polym. Sci., Polym. Chem. Ed. *14*, 83 (1976); Banihashemi, A., Kiaizadeh, F.: Makromol. Chem. *181*, 325 (1980)
22. Babé, S. G., de Abajo, J.: Rev. Plasticos Modernos *22* [182], 1 (1971)
23. Wrasidlo, W., Levine, H. H.: J. Polym. Sci., Part A *2*, 4795 (1964)
24. Gerber, A. H.: J. Polym. Sci., Polym. Chem. Ed. *11*, 1703 (1973); see also Gerber, A. H.: US Pat. 3,943,125 (1976); Chem. Abstr.: *85*, 33936 (1976)
25. (a) Korshak, V. V. et al.: Macromolecules *5*, 807 (1972); (b) Tugushi, D. S. et al.: Vysokomol. Soedin., Ser. A *15*, 969 (1973); (c) Tugushi, D. S. et al.: Tr. Mosk. Khim.-Tekhnol. Inst. [70], 182 (1972); Chem. Abstr.: *79*, 19161 (1973)
26. Pravednikov, A. N. et al.: Plaste Kautsch. *22*, 476 (1975)
27. Zurakowska-Orszagh, J., Kurzela, M., Kaminski, J.: Polimery (Warsaw), *25*, 51 (1980); Chem. Abstr.: *93*, 168651 (1980)
28. Korshak, V. V. et al.: Vysokomol. Soedin. Ser. A *21*, 122 (1979); Korshak, V. V. et al.: USSR Pat. 652,193 (1979); Chem. Abstr. *90*, 205176 (1979); Korshak, V. V. et al.: Izv. Akad. Nauk Gruz. SSR, Ser. Khim. *6*, 122 (1980); Chem. Abstr. *94*, 66132 (1981); Korshak, V. V., Rusanov, A. L., Tugushi, D. S.: USSR Pat. 749,859 (1980); Chem. Abstr. *94*, 16350 (1981)

29. Korshak, V. V. et al.: Soobshch. Akad. Nauk Gruz. SSR *96,* 341 (1979); Chem. Abstr. *92,* 164327 (1980)
30. Iwakura, Y., Uno, K., Imai, Y.: J. Polym. Sci., Part A *2,* 2605 (1964)
31. Imai, Y., Uno, K., Iwakura, Y.: Makromol. Chem. *83,* 179 (1965)
32. Srinivasan, P. R., Mahadevan, V., Srinivasan, M.: Makromol. Chem. *180,* 1845 (1979)
33. Iwakura, Y., Uno, K., Chau, N.: Makromol. Chem. *176,* 23 (1975)
34. Neuse, E. W., Horlbeck, G.: Polym. Eng. Sci. *17,* 821 (1977)
35. Korshak, V. V. et al.: Vysokomol. Soedin, Ser. A *18,* 2585 (1976)
36. Varma, I. K., Veena: J. Polym. Sci., Polym. Chem. Ed. *14,* 973 (1976); J. Macromol. Sci., Chem. *11,* 845 (1977)
37. Sivriev, H., Borissov, G.: Eur. Polym. J. *13,* 25 (1977)
38. Korshak, V. V. et al.: Vysokomol. Soedin., Ser. A *22,* 1209 (1980)
39. Kojima, T.: J. Polym. Sci., Polym. Phys. Ed. *18,* 1685 (1980)
40. Banihashemi, A., Fabro, D., Marvel, C. S.: J. Polym. Sci. Part A-1, *7,* 2293 (1969); Banihashemi, A., Marvel, C. S.: Iranian J. Sci. Tech. *2,* 203 (1972)
41. Marvel, C. S.: Techn. Rep. ML-TDR-64-39, Part I (1964)
42. Kokelenberg, H., Marvel, C. S.: J. Polym. Sci., Part A-1, *8,* 3199 (1970)
43. Hedberg, F. L., Marvel, C. S.: J. Polym. Sci., Polym. Chem. Ed. *12,* 1823 (1974); Nuova Chim. *50,* 51 (1974)
44. Dudgeon, C. D., Vogl, O.: J. Polym. Sci., Polym. Chem. Ed. *16,* 1815, 1831 (1978); Dudgeon, C. D.: PhD Thesis, Univers. of Massachusetts (1976). A related, low-molecular-mass polymer containing the purine ring system was prepared by the same basic method: Dudgeon, C. D., Vogl, O.: J. Macromol. Sci. Chem. *11,* 1989 (1977)
45. Gray, D. N., Rouch, L. L., Strauss, E. L.: Polym. Prepr. *8,* 1138 (1967)
46. Higgins, J., Marvel, C. S.: J. Polym. Sci., Part A-1, *8,* 171 (1970)
47. D'Alelio, G. F.: US t. 3,763,107 (1963); Chem. Abstr.: *80,* 71350 (1974)
48. Neuse, E. W.: Chem. Ind. (London), *1975,* 315
49. Unpublished results from the author's laboratory
50. Coville, N. J., Neuse, E. W.: J. Org. Chem. *42,* 3485 (1977)
51. Although not as yet demonstrated for the polymer case, H_2O_2 has been shown to be the by-product in analogous nonpolymeric reactions: Grellmann, K. H., Tauer, E.: J. Am. Chem. Soc. *95,* 3104 (1973)
52. The method was first employed by Grellmann and Tauer (preceding ref.) in a study of certain *o*-substituted benzylideneanilines
53. Iwakura, Y. et al.: Makromol. Chem. *77,* 41 (1964)
54. Korshak, V. V. et al.: Rev. Roum. Chim. *22,* 1521 (1977)
55. Hara, S., Senoo, M., Taketani, Y.: Japan. Kokai *1974,* 74 78,798
56. Manecke, G., Brandt, L., Kossmehl, G.: Makromol. Chem. *178,* 1745 (1977)
57. Denyes, R. O.: Techn. Rep. AFML-TR-66-167, Vols. I–III (1966)
58. Marvel, C. S.: J. Macromol. Sci. – Revs. Macromol. Chem. *13,* 219 (1975)
59. Covacs, H. N., Delman, A. D., Simms, B. B.: J. Polym. Sci., Part A-1, *6,* 2103 (1968)
60. Izyneev, A. A. et al.: Dokl. Akad. Nauk SSSR *231,* 1126 (1976)
61. Osgerby, J. M., Pauson, P. L.: J. Chem. Soc., 4604 (1961); Omote, Y., Kobayashi, R., Nakara, Y., Sugiyama, N.: Bull. Chem. Soc. Jpn. *46,* 3315 (1973)
62. Kojima, T.: Polym. J. *13,* 85 (1981)
63. Kojima, T. et al.: J. Polym. Sci., Polym. Phys. Ed. *18,* 1673 (1980)
64. Ryan, M. T., Helminiak, T. E.: Polym. Prepr., Am. Chem. Soc. Div. Polym. Chem. *14,* 1317 (1973)
65. Kojima, T.: J. Polym. Sci., Polym. Phys. Ed. *18,* 1791 (1980)
66. Belohlav, L. R.: Angew. Makromol. Chem. *40/41,* 465 (1974)
67. Leal, J. R.: Mod. Plastics *52,* 60 (1975)
68. Levine, H. H., Stacy, R. D.: Techn. Rep. AFML-TR-65-350 (1966)
69. Gibbs, W. E., Helminiak, T. E.: Polymers for Use at High and Low Temperatures in: Polymer Science, Vol. 2, North-Holland Publishing, New York 1972
70. Slade, P. E., Jenkins, L. T.: Thermal Characterization Techniques, Vols. 1 and 2, Marcel Dekker, New York 1966 and 1970
71. Neuse, E. W.: J. Organomet. Chem. *6,* 92 (1966)

72. Shulman, G. P., Lochte, W.: Polymer Prepr. *6,* 773 (1965); J. Macromol. Sci. Chem. *1,* 413 (1967). For more recent, complementary results see Chatfield, D. A., Einhorn, I. N.: J. Polym. Sci., Polym. Chem. Ed. *19,* 601 (1981)
73. Dubey, G., Kane, J. J.: Am. Chem. Soc., Div. Org. Coat. Plast. Chem. Papers *30,* 187 (1970)
74. Reed, R., Hidde, R.: Techn. Rep. AFML-TR-65-146, Vols. I and II (1965)
75. Levine, H. H., Delano, C. B., Kjoller, K. J.: Polym. Prepr. *5,* 160 (1964)
76. Marks, B. S., Shoff, L. E., Watsey, G. W.: Techn. Rep. NASA CR-1723 (1971)
77. Kalnin, I. L., Breckenridge, G. J.: US Pat. 4,113,683 (1978); Chem. Abstr.: *90,* 24256 (1979)
78. Wirsen, D.: Kem. Tidskr. *91,* 16 (1979)
79. Matsushita Electric Works, Ltd., Jpn. Tokyo Koho, Jap. Pat. 80 28,438 (1980); Chem. Abstr.: *94,* 23967 (1981)
80. Minges, M. L.: High Temp. – High Pressures *1,* 607 (1969). Marks, B. S., Rubin, L.: J. Macromol. Sci. Chem. *3,* 555 (1969); Dickey, R. R., Lundell, J. H., Parker, J. A.: Macromol. Sci., Chem. *3,* 573 (1969); D'Alelio, G. F., Parker, J. A.: Ablative Plastics, Marcel Dekker, New York 1971
81. Lipskis, A. et al.: Liet. TSR Mokslu Akad. Darb., Ser. B *1972,* 159; Chem. Abstr.: *79,* 92894 (1973)
82. Dunay, M.: US Pat. 3,775,213 (1973); Chem. Abstr.: *80*, 71793 (1974). Sheratte, M. B.: US Pat. 4,154,919 (1979); Chem. Abstr.: *91*, 92439 (1979).
83. Verzwyvelt, S. A.: Fr. Dem. 2440084 (1978); Chem. Abstr. *94*, 124660 (1981). Martin, R. E.: Technical Rep. NASA-CR-159653, FCR-1017 (1978); Yaffe, M. R., Murray, J. N.: Proc. DOE Chem./Hydrogen Energy Syst. Contract. Rev. (CONF-781142), 37 (1978; publ. 1979); Chem. Abstr. *94,* 22105 (1981)
84. Reed, R., Hidde, R.: J. Polym. Sci., Part A *1,* 1531 (1963); Technical Report AFML-TR-65-136 (1965)
85. Hergenrother, P. M.: SAMPE Quarterly *3* [1], 1 (1971)
86. Gosnell, R. B., Levine, H. H.: J. Macromol. Sci. Chem. *3,* 1381 (1969)
87. Ross, J. H., Opt, P. C.: Appl. Polym. Symp. [9], 23 (1969), and earlier reports from this group; Singleton, R. W.: Appl. Polym. Symp. [9] 133 (1969)
88. Conciatori, A. B., Smart, C. L.: US Pat. 3,502,606 (1970); Chem. Abstr. *72,* 112737 (1970). Bohrer, T. C., Rosenthal, A. J.: US Pat. 3,502,756 (1970); Chem. Abstr. *72,* 112727 (1970)
89. Tesoro, G., Moussa, A.: Fire Retard., Proc. Eur. Conf. Flammability, Fire Retard., 2nd, 159 (1978; pub. 1980); Chem. Abstr. *92,* 182415 (1980)
90. Defosse, T. C., Welch, I. H.: Mod. Text. *52,* 65 (1971)
91. Chenevey, E. C.: Technical Rep. NASA-CR-152281 (1978); Technical Report NASA-CR-159644 (1979)
92. Kalnin, I. L., Powers, E. J.: US Pat. 3,903,248 (1975); Chem. Abstr. *83,* 195038 (1975)
93. Ezekiel, H. M., Spain, R. G.: US Pat. 3,528,774 (1970); Chem. Abstr. *73,* 110796 (1970). Stuetz, D. E.: Am. Chem. Soc., Div. Org. Coat. Plast. Chem. Papers *31,* 389–99 (1971)
94. Model, F. S., Lee, L. A.: Polybenzimidazole Reverse Osmosis Membranes, in: Reverse Osmosis Membrane Research, Lonsdale, H. K., and Podall, H. E., Eds., Plenum Publishing Co., New York 1972
95. Tan, M., Davis, H. J.: Techn. Rep. W 79-07571, OWRT-7527(1) (1978); Chem. Abstr. *92,* 99356 (1980); Goldsmith, R. L. et al.: Technical Rep. W 79-09314, OWRT-7509(1) (1979); Chem. Abstr. *92,* 203270 (1980). Senoo, M., Mori, K., Taketani, Y.: Ger. Pat. 2,559,931 (1979); Chem. Abstr. *92,* 111507 (1980)
96. Suzuki, H. et al.: Bull. Chem. Soc. Jpn. *48,* 1922 (1975)
97. Rodé, V. V. et al.: Izv. Akad. Nauk SSSR, Ser. Khim. *1968,* 2662
98. Le Maux, P. et al.: J. Org. Chem. *45,* 4524 (1980), and refs. cited therein

Received November 2, 1981
W. Kern (editor)

Some Problems Encountered with Degradation Mechanisms of Addition Polymers

A. L. Bhuiyan

Department of Chemistry, The University of The West Indies, St. Augustine, Trinidad, West Indies

It is shown in this article that such parameters as intramolecular cyclization, steric effects, and resonance stabilization can most satisfactorily explain the yields of monomers in thermal degradation of addition polymers.

Advances in Polymer Science 47

I. Introduction

Studies on thermal degradation of high polymers probably started with the thermal degradation of natural rubber[1], followed by the investigation in the degradation of starch by Meyer et al.[2] and of cellulose by Kuhn[3]. Kuhn has shown that there is a close agreement between theory and experiment in the initial stages of degradation if every hydrolyzable bond in the cellulose molecule is considered to be equally easily scissioned.

Later, Kuhn and Freudenberg[4], in order to account for the course of the later stages of the degradation, have suggested that hydrolysis of the bonds in the biose and triose residues occurs more rapidly than in the polysaccharide and that all terminal bonds break more quickly than those along the chain. Simha[5] has discussed in a theoretical treatment the possibility of obtaining information as to whether the degradation occurs by a mechanism generally similar to that for polysaccharides or whether preferred scission at the polymer chain-ends may occur. Montroll and Simha[6] have calculated statistically the changes in average chain length and in the distribution of chain lengths, based on the assumption that all bonds are equally likely to be scissioned. Schulz[7] has studied the hydrolysis of nitrocellulose and has assumed that it has an equal number of weak links placed at equal distances along the chain. He justifies the assumption by the statement that if the weak links are placed at random over each chain, distributions identical with those for chains without weak links are obtained. Some authors[8–11] have observed methyl methacrylate as the predominant product of degradation of poly(methyl methacrylate). Votinov et al.[8] claim to have obtained good agreement between theory and experiment by applying Kuhn's theories to the degradation of poly(methyl methacrylate). The authors have observed the evolution of large quantities of monomer right at the beginning of the reaction. However, Shimizu and Munson[11] have observed traces of such other minor products as dimer and tetramer of methyl methacrylate, ethyl methacrylate, ethyl butanoate, and perhaps methyl butanoate in addition to the major product methyl methacrylate[9, 10]. McNeill et al.[12] have analyzed the degradation products of poly(methyl methacrylate) under vacuum at 500 °C. The TVA curve for degradation products of poly(methyl methacrylate) shows a single peak corresponding to methyl methacrylate as the only volatile product. Grassie and Melville[13] have studied the degradation of poly(methyl methacrylate) in the temperature range of 170°–260 °C in vacuum and observed that, for polymer of molecular weight 44300, there is no fall in molecular weight up to at least 65% degradation and the curve of molecular weight versus percent degradation to monomer follows the path AB (Fig. 1). But the curves for the higher molecular weights polymers start falling away from the path AB till with molecular weights 650000 and 725000, the degradation follows the curve AC, indicating that the probability of complete disintegration to monomer decreases with the increase in molecular weight, which contradicts the theory. The deviation is attributed to the probability of termination being the same as that for concurrent evolution of monomer. The authors have offered two explanations for the experimental results. These are:

(a) fission occurs at random along the polymer chain, elimination of monomer ensues, but the two polymer radicals combine before complete disintegration, and

(b) fission occurs at the chain-ends exclusively, elimination of monomer occurs from the radicals and the polymer radicals are eventually stabilized by disproportionation reaction.

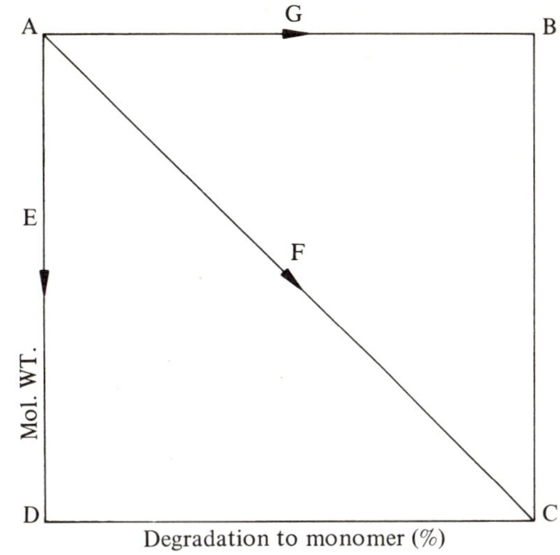

Fig. 1

In other words, *one molecule of the monomer would be peeled off from the ends of the chains of the higher molecular weight polymers and the decrease in molecular weight would be proportional to the number of monomer molecules evolved.* The same number of average molecular weight would result whether one molecule is liberated from each of n polymer molecules or whether n monomers are liberated from one polymer molecule. Therefore, the main reaction in 650 000 and 725 000 polymers is in fact reverse polymerization but every degrading chain is established before it disappears completely and becomes once more a stable molecule, but of shorter chain length.

The same authors[14)] have also studied the effects of particle size on the rate of production of monomer during thermal degradation of poly(methyl methacrylate) and observed that the rate increases with particle size for particles smaller than about 50 mesh but it does fall away for larger particles. The study of the effect of layer thickness on the rate of monomer production has also revealed the similar pattern of behavior of the polymer towards degradation. Several reasons for this behavior of the degradation are mentioned. These are:

(a) the monomer molecules are trapped inside the polymer, and
(b) the larger particles are not uniformly heated, because a polymer is a bad conductor of heat and the side of the particles away from the tray might be at a temperature considerably lower than the experimental temperature (260°).

Their later studies[15)] on thermal degradation of the same polymer in presence of an inhibitor have led to the conclusion that the mechanism is *a chain reaction involving free radicals*. The double and single bonds ends break most easily and, therefore, the scission occurs initially at the chain-ends. This is supported by the fact that no difference between the viscosity of a solution of the original polymer and a solution of the polymer degraded to the extent of 27% in presence of the inhibitor has been observed. Cowley and Melville[16)] have suggested that the termination of the chain radicals formed in the degradation of poly(methyl methacrylate) is diffusion-controlled. The authors have further stated that in the photo-initiated thermal degradation of the same polymer there is a considerable transfer of hydrogen. While Grassie and Melville[14)] interpret their results on thermal degradation of poly(methyl methacrylate) as an end initiation followed by depropagation which consumes completely the radicals of small N; the radicals of larger N having long enough life times to terminate bimolecularly, Brown and Wall[17] attribute the observed dependence of initial rate on molecular weight of the same polymer to transfer. Bywater and Black[18] have investigated the thermal degradation of poly(methyl methacrylate) in solution in 1,2,4-trichlorobenzene and α-methylnaphtalene. The experimental results show that the mechanism involving initiation at the chain-ends, depropagation, and chain breaking by solvent transfer holds for all the solvents studied. Recently, several conflicting reports[19–23] on kinetics and degradation mechanism of poly(methyl methacrylate) have appeared in the literature. The disagreement between the results of these works seems to spread over a large number of factors which are believed to influence the degradation of the polymer. These factors are initial molecular weight[20, 22, 23], range of temperature used[20–23], morphology of the polymer[22], and the presence of double-bond ends[19]. However, it is suggested[21, 23] that the depropagating chains are predominantly end-initiated and the chain termination is bimolecular, at the lowest temperatures used (about 300 °C), but at the highest temperatures (about 500 °C), the initiation is caused by predominantly random chain scission, most of the chains are terminated as the depropagation reaches the end of the chain and the terminal radical then distils out the system.

Reich and Stivala[24] have deduced the following expression [Eq. (1)] connecting the ratio of the weight (W) of the polymer at a certain instant to the weight (W_0) before degradation and conversion.

$$W/W_0 = 1 - C = B^{b/(b-1)} \tag{1}$$

where $B = D_p/D_{p,0}$ and C = conversion = $(W_0 - W)/W_0$.

The equation (1) holds for fractionated sample (b = 1) of poly(methyl methacrylate) of molecular weight ($M_{n,0}$) 30000 to 40000 at the high temperatures at which the molecular weight does not change. For $b > 1$, a decrease in molecular weight has been observed at 500 °C in agreement with the theory. The results agree with the degradation of lower molecular weight samples in which the degradation is suggested to be initiated by random chain scission followed by complete depropagation. But with high molecular weight samples, decrease in the M_n with conversion is observed. The authors have offered several explanations:

(a) due to imperfection in the fractionation procedure the high-molecular-weight samples may be sufficiently heterodisperse to exhibit some characteristics of unfractionated samples;
(b) the sample may pass through low and intermediate temperatures for most part of the very short reaction time so that some of the characteristics of the degradation mechanism corresponding to these temperatures are exhibited; and
(c) there is still a finite probability of bimolecular termination, at the high temperature, which may become significant if the initial molecular weight of the fraction exceeds the mean kinetic chain length.

It seems that the method of preparation of a polymer has significant influence on its mode of degradation[25]. Thus, a sample of poly(methyl methacrylate), prepared by a free radical reaction, suffers rapid depolymerization at about 275 °C and a second mode of initiation of chain depolymerization between 350 and 400 °C, while the sample, prepared by anionic polymerization, undergoes depolymerization as a whole above 350 °C. Absence of any degradation of the anionically prepared polymer below 350 °C is attributed to the absence of double bonds at the end of the chain, while due to the presence of double bonds at the chain-ends of the polymer sample, prepared by free radical reaction, the degradation is initiated at lower temperatures at the chain-ends[26]. According to Cameron and Kerr[27], the chain-end initiation must be included in the expressions for kinetics in order to describe fully the rate of monomer formation, provided the zip length is less than the mean molecular chain length. However, Davis[28] observes that the degradation of the irradiated poly(methyl methacrylate) occurs via chain-end diffusion and unzipping.

Poly(α-methylstyrene) undergoes rapid photo-degradation to volatile products in the range 240–270 °C, where there is also an appreciable rate of thermal degradation[16]. Brown and Wall[17] measured the rates of volatilization and decrease in the degree of polymerization of poly(α-methylstyrene) at 282 °C in vacuum. The rates of volatilization increase with initial DP, while the DP of the residue decreases moderately. The rate of pyrolysis decreases monotonously with conversion and then levels off at high conversion (about 60–70%). No maximum in the rate versus conversion curve is observed. The DP is also found to decrease with conversion and the decrease observed with poly(α-methylstyrene) is more than that observed with poly(methyl methacrylate) but less than that with polystyrene. The experimental results obtained from 97000 molecular weight poly(α-methylstyrene) do fit closely into the theoretical curves constructed for the case of random initiation but those from higher molecular weights remain far outside. The results have been interpreted in terms of following chain-reaction degradation mechanism, which has been developed in a quantitative fashion by Simha, Wall, and Blatz[29–31].

Initiation:

$$\sim CH_2CHXCH_2CHX\sim \rightarrow \sim CH_2\dot{C}HX + \dot{C}H_2CHX\sim \tag{2}$$

Propagation:

$$\sim CH_2CHXCH_2\dot{C}HX \rightarrow \sim CH_2\dot{C}HX + CH_2{=}CHX \tag{3}$$

Intermolecular transfer:

$$\sim CH_2\dot{C}HX + \sim CH_2CHXCH_2CHX$$

$$\rightarrow \sim CH_2CH_2X + \sim CH_2CX{=}CH_2 + \dot{C}HX\sim \tag{4}$$

Intramolecular transfer:

$$\sim CHXCH_2CHXCH_2\dot{C}HX$$

$$\rightarrow \sim \dot{C}HX + CH_2{=}CXCH_2CH_2X \tag{5}$$

Termination:

$$2R \rightarrow 2P \tag{6}$$

Depolymerization of poly(α-methylstyrene) has been studied also in solution in various solvents[32–35]. The results have been interpreted in terms of a chain reaction initiated by random scission, followed by a rapid depropagation, no transfer and a kinetic chain length high compared with the molecular chain length of the polymer. Chain initiation by the scission of weak or abnormal links is ruled out on the grounds that the rates of depolymerization of cationically and anionically produced polymers are closely similar. The results are in agreement with those reported by Brown and Wall[17)] who studied the degradation of the molten polymer, except that there is a difference between the rates of production of monomer. At a temperature of 231° and for a polymer of molecular weight 350000, the rate of monomer production in bulk degradation experiments[17, 36)] should correspond to 40–100 μmol/ml h. But this rate seems to be low compared to the solution rate described[33)] as 22 μmol/ml h at one-twentieth the polymer concentration. Part of the discrepancy is attributed to the lower kinetic chain length in bulk polymer. But Jellinek and Luh[35)] have found the rate of monomer production strongly dependent on the viscosity of the solvent, indicating that the rate is diffusion-controlled.

The early investigation of thermal degradation of polystyrene has been carried out by Staudinger and Steinhofer[37)] in order to account for the formation of dimer and trimer and to determine the head-to-tail arrangement of the repeating units in this polymer.

Jellinek[38)] has investigated thermal degradation of polystyrene in vacuum at about 300 °C and observed some discrepancies in the theory developed by Kuhn, Mark, Schulz, Montroll, and Simha. The main assumptions of their theory are:

(a) each link in a polymer chain has equal strength and equal probability of scission,
(b) scission of links occurs at random, and
(c) the rate of scission is proportional to the number of links present.

The major discrepancies found by Jellinek are:

(a) degradation stops completely or slows down markedly when a certain chain length is reached,
(b) the distribution curves are narrower than those predicted by theory, and
(c) the quantity of monomer production during degradation is about 10^3 to 10^4 times larger than expected.

The author deduced a complex expression for statistical distribution of weak links at random over each chain and calculated the number and weight distribution functions as well as weight average chain length. It has been shown that the degradation in the case of the chains containing 5 weak links stops at a weight average chain length of 286 units, whereas normal chains degrade completely to monomer. Later, the same author[39)], in dealing with the degradation of long-chain molecules, in which the normal as well as the weak links would undergo scission, each with a separate rate constant, has stated that this type of degradation would occur in three steps:

(a) *initiation:* formation of two active chain-ends by the initial scission of a weak link,
(b) *propagation:* peeling-off of monomer units from these chain-ends, and
(c) *termination:* deactivation of the chain-ends thus terminating the liberation of monomer units.

Shimizu and Munson[11)] have carried out pyrolysis/mass spectrometric studies of polystyrene at 370–390 °C and have found styrene, dimers, and trimers as the major products and C_9H_8, C_9H_{10}, $C_{15}H_{14}$, and $C_{18}H_{18}$ as the minor products. The ratio of styrene: dimer: trimer has been found to be 1 : 0.4 : 0.2. McNeill et al.[12)] have observed only styrene as the major product of pyrolysis of polystyrene at 380 °C under vacuum. While Jellinek[40)] suggests that the rapid decrease in molecular weight is the result of scission of weak points in the polymer chain, Hall[41] maintains that polystyrene degrades first by rapid chain scission to shorter fragments which in turn degrade stepwise to monomer. Mechanism of thermal degradation of polystyrene has been suggested to involve three reactions:

(a) depolymerization by depropagation initiated at the chain-ends,
(b) random scission of the larger molecules, and
(c) transfer reactions, intramolecular and/or intermolecular. Though Jellinek's weak points are now believed to be the sites of occasional unsaturated linkages which are formed during polymerization[42], according to Grassie and Kerr[43], some of these weak points could be due to occasional head-to-head addition of styrene units. The experimental results obtained by Grassie and Kerr[43] seem to satisfy the theoretical expressions deduced by Gordon[44] on the basis of chain-end initiation, intermolecular transfer, and bimolecular termination. Investigations of thermal degradation of deuteropolystyrene have shown that the monomer yield for this polymer is about 70% by weight, while it is about 42% for the undeuterated polymer and that the initial molecular weight decrease is reduced by about 50% by deuterium substitution[45, 46]. This is ascribed to the isotope effect in intramolecular transfer. The results of these investigations are, therefore, in agreement with the weak link theory developed by Jellinek. Gordon's mechanism, which fits the experimental results obtained by Grassie and Kerr[43], is in conflict with the results of investigations of degradation of polystyrene in bulk and in naphthalene or tetralin solution[47, 48] in which tetralin has been found to inhibit the volatile formation, while the initial decrease in molecular weight remains unchanged. Rate of formation of volatile products has been found to be proportional to the number of chain ends[49], which has been confirmed by the works of Cameron[50]. Also, these investigations indicate that the formation of the volatile products is a free radical process which is initiated at the chain ends and that the dimers, trimers, tetramers, etc., are formed by intramolecular transfer. Cameron and McWalter[51,52] have attributed the sharp decrease in molecular weight in thermal degradation of polystyrene to intermolecular transfer. The authors also maintain that the intramolecular transfer, and not the intermolecular transfer, is mainly responsible for the production of volatile oligomers, such as, dimers, trimers, tetramers, etc. The absence of weak links in polystyrene has been revealed by the investigations of thermal degradation of non-radical polystyrene by several workers[53–56].

Investigations on thermal degradation of polyethylenes, linear and branched, have been carried out by a number of workers[57–75]. Both the linear and branched polyethylenes yield, on pyrolysis, less than 1% monomer.

Madorsky et al.[67] obtained, besides negligible quantities of ethylene, a wide variety of linear hydrocarbons consisting mainly of a waxlike fraction of average molecular weight 696. Initiation by random scission or at weak links has been proposed[57]. The products of low molecular weight are formed by inter- or intramolecular free radical transfer. A free radical mechanism has been evoked[57, 60–62] for explaining the formation of various products. The mechanism is given below:
Initiation: scission of weak bonds or ordinary C–C bonds.

$$R\text{–}R \rightarrow 2\,R^{\cdot} \tag{7}$$

Propagation:

$$R^{\bullet} \rightarrow R_1^{\bullet} + CH_2{=}CH_2 \qquad (8)$$

Intramolecular transfer followed by degradation.

$$R_1^{\bullet} \longrightarrow R_2{-}\dot{C}H{-}CH_2{-}R_3 \longrightarrow R_2^{\bullet} + CH_2{=}CH{-}CH_2{-}R_3 \qquad (9)$$

$$\phantom{R_1^{\bullet} \longrightarrow R_2{-}\dot{C}H{-}CH_2{-}R_3} \longrightarrow R_2{-}CH{=}CH_2 + R_3^{\bullet} \qquad (10)$$

Intermolecular transfer followed by degradation.

$$R_4{-}CH_2{-}R_5 + R^{\bullet} \rightarrow R_4{-}\dot{C}H{-}R_5 + RH \qquad (11)$$

$$R_4{-}\dot{C}H{-}R_5 \rightarrow R_4{-}CH{=}CH_2 + R_6^{\bullet} \qquad (12)$$

Termination:

$$R_n^{\bullet} + R_m^{\bullet} \rightarrow R{-}CH{=}CH_2 + R_mH \qquad (13)$$

$$R_n^{\bullet} + R_m^{\bullet} \rightarrow R_n{-}R_m \qquad (14)$$

The formation of the most abundant degradation products, such as, propane and 1-hexane has been explained by intramolecular transfer via a six-membered cyclic ring structure. Oakes and Richards[63)] have put forward a dual mechanism for thermal degradation of polyethylene, a fast reaction through tertiary hydrogen atoms and a slow one through secondary hydrogen atoms. The degradation products are plastics or waxlike materials, or liquids. Gaseous products, of which ethylene is a minor constituent, are not formed below 370 °C. Three types of olefinic double bonds, such as $RCH{=}CH_2$, $RCH{=}CHR'$, and $RR'C{=}CH_2$, identified by infrared analysis, are formed during the pyrolysis of branched polyethylene by hydrogen transfer. Polymer with less branching gives higher proportion of vinyl groups. The increase of vinyl groups with increase in the degree of degradation has been attributed to the preferential abstraction of tertiary hydrogen present at the branching points by the radicals. The molecular weight of branched polyethylene has been found to decrease above 290 °C, but the weight loss is negligible up to 370 °C, a result confirmed by Wall et al.[64)]. Thermal degradation of linear and branched polyethylene has been investigated by Wall and Straus[58)]. The linear polyethylene obeys the theory of random degradation, but the branched material does not. Rate of degradation increases with increase in branching and the greater the branching, the greater the deviation from random theory. Madorsky[68)] interprets his findings with the help of theoretical deductions based on a free radical chain reaction involving initiation, propagation, transfer, and termination[29–31)]. But the rate versus conversion curve for branched polyethylene does not have a maximum at 25%, suggesting the absence of transfer and random scission. Igarashi and Kambe[59)] have studied thermal degradation of low and high pressure polyethylene in nitrogen and have observed that the high pressure polyethylene starts decomposing at a higher temperature (20 °C) than the low pressure material. Activation energies of 61–74 kcal are obtained. Several authors have detected fragments up to C_{70}[75)], C_{50}[74)], C_{55}[73)], and C_{13}[76)] in the pyrolysis of polyethylene and C_{60}[73)] in the cases of polypropylene and polyisobutylene. A random cleavage process has been concluded[74)] from the slope of the distribution above C_{11}. It is surprising to note that the final temperature of the pyrolyzer (between 550 and 1000 °C) has no influence on the probability of random cleavages, as calculated from the distribution[77]. This indicates that the vaporization, which favors stable long fragments, is accelerated with increasing temperature. Based on the experimental results, a mechanism of random scission followed by volatilization has been postulated[73] for degradation of linear polyethylene, polypropylene, and polyisobutylene. Seeger and Barrall[72] have studied pyrolysis-gas chromatographic analysis of chain branching in polyethylene. The authors have made the following observations:

(a) primary random scission is followed by stabilization by hydrogen transfer and

(b) decomposition occurs by an intramolecular cyclization process via transfer of radicals from the first to the fifth carbon followed by a chain cleavage at the β-position.

An increased probability of scission at the α- and β-positions to a tertiary carbon atom has also been observed in contrast to a linear polyethylene chain[78]. Seeger and Barrell[72] further maintain that the composition and yield of products are equipment-dependent. The diversity of pyrolysis equipment, as employed by many workers, has further complicated the general application of the technique and programs made in one laboratory have usually not been reproducible directly in another laboratory. Rideal and Padget[71] have observed decrease in melt viscosity and a narrowing of molecular weight distribution at high melting temperatures (>290 °C) and increase in melt viscosity at lower melt temperatures. The increase in melt viscosity has been explained by molecular enlargement due to the formation of long chain branches. It has been observed that the scission and enlargement reactions are not mutually exclusive but competitive. Heating samples at temperatures between 284 and 355 °C under high purity grade nitrogen has revealed that the high density polyethylene does not degrade according to a random scission type of mechanism[70]. Instead both molecular diminishing and enlargement have been found to occur simultaneously. Jellinek[69] has attributed the production of the very small amount of monomer in the degradation of polyethylene to the presence of labile hydrogen atoms. Polymers containing no labile hydrogen atoms, for instance, polymers containing α-deuterium atom or an α-methyl group produce higher amount of monomers. But this does not explain the formation of only 32 weight % of monomer in thermal degradation of polyisobutylene (Table 1), because, due to the presence of α-methyl group, a very high yield would have been expected.

Monomers yields for some chain-growth olefin polymers on thermal degradation at 300–500 °C in vacuum[32, 76, 77] are given in Table 1 for comparison.

In the above review two extreme categories of polyolefins have been described. Poly(methyl methacrylate) and poly(α-methylstyrene), which yield about 100% monomer on thermal degradation, form one extreme, while polyethylene, which yields less than 1% monomer, forms the other extreme. Polystyrene, which yields about 40% monomer, exhibits an intermediate behavior. It is evident from the review that there is large difference in monomer yields on thermal degradation of various polymers under identical conditions and that for a certain polymer the experimental results are similar in various studies, but no completely satisfactory interpretations have been given, capable of explaining all the features of the degradation, some of which appear to be contradictory. It is the object of this paper to attempt to get a much closer insight into some of the parameters which influence the yield of monomers in thermal degradation of polyolefins and identify those which are most important and largely responsible for large differences in monomer yields. Intramolecular cyclization has been discussed by Mita[81] on the basis

Table 1

Polymer	Monomer yield, Wt. %
Polyacrylate	
Methy acrylate	0.7
Methyl methacrylate	92–100
Polystyrene	
styrene	42
α-Deuterostyrene	70
β-Deuterostyrene	42
α-Methylstyrene	95–100
α,β,β-Trifluorostyrene	74
Polyhydrocarbon	
Ethylene (linear and branched)	0.03
Propylene	0.2–2.0
Isobutylene	18–32
Polyfluorocarbon	
Tetrafluoroethylene	97–100
Chlorotrifluoroethylene	28

of the principle of "backbiting". The same topic will be discussed below in this paper on the basis of crystal morphology.

II. Parameters

A. Zip Length

The expression for zip length is given by

$$\text{Zip length} = \frac{\text{Probability of propagation}}{\text{Probability of (transfer + termination)}}$$

This expression implies that with unusually high zip length the probability of propagation would be very high and that of transfer and termination very low. On the other hand, if the zip length is very small or negligible, then there would be negligible propagation but transfer and termination would be very high. In the former case depolymerization would dominate over transfer and termination and the whole of the polymer chain would almost completely unzip into monomer. In the latter case, mainly non-monomeric products would be formed. Poly(methyl methacrylate) has been found to produce, on thermal degradation, 92–100% monomer. The zip length for this polymer has been calculated[82] from the kinetic studies to be 60. But the zip length of a poly(methyl methacrylate) polymer of molecular weight 44300 which would volatilize to the extent of 65% of its initial weight without change in the molecular weight of the residue should be about 400[83]. Recently, the zip length for the same polymer has been reported[84] to be of the order of $10^2 - 10^4$. The residue in thermal degradation of polyethylene[63, 66] at

300–500 °C shows a thousand-fold decrease in molecular weight after the polymer has volatilized to the extent of only 2%, suggesting that the zip length for this polymer is very small. Actually, the calculated[84)] value for zip length of polyethylene at 400 °C is of the order of $10^{-1} - 10^{-2}$. This means that the propagation is negligible as shown by the production of less than 1% monomer. Transfer and termination are, therefore, the dominant parameters. Although polyethylenes, both linear and branched, produce less than 1% monomer and Wall et al.[64)] have observed little effect of branching on the rate of depolymerization, there is no maximum in the rate versus conversion curve for branched polyethylene. The absence of a maximum indicates that random initiation and transfer are absent and that the continuous rate decrease is similar to that resulting from a reverse propagation mechanism. A maximum in the rate versus conversion curve for polystyrene[37, 85)] at a conversion rate of about 40% rather than 25% as predicted by the theory for random degradation suggests that the process is non-random but involves considerable transfer. The marked increase in the monomer yield on thermal degradation in going from polystyrene to poly(α,β,β-trifluorostyrene) through poly(α-deuterostyrene) and poly(α-methylstyrene) in which the C–H bond at the α-position is replaced by a C–D, or C–CH_3, or C–F bond indicates that intermolecular and/or intramolecular transfer is responsible for the formation of non-monomeric products in thermal degradation of polystyrene. Based on these results, it has been suggested that the chain-transfer site in polystyrene is exclusively the benzylic α-carbon-hydrogen bond of the repeating unit. The relatively high yields of dimers (19%) and trimers (23%) obtained from polystyrene also indicate the prevalence of transfer processes in the degradation reaction. If the α-carbon-hydrogen bond of the repeating unit is exclusively the transfer site, then the number of hydrogen atoms abstracted from the β-position should be negligible compared to that abstracted from the α-position. Studies[86)] on the γ-irradiation of polystyrene, deuterated as well as undeuterated, have revealed that the ratio between the number of hydrogen atoms abstracted from the α-position and that from the β-position is approximately unity for the same amount of energy spent, indicating that the case of abstraction of hydrogen atoms at α- and β-positions does not differ appreciably. The G-values for C–H and C–D bonds and the resulting isotope effects are given in Table 2.

The authors[86)] have further observed that when a deuterium atom is attached to the same carbon atom as a hydrogen, the C–H bond is broken 2.3 times as readily as a C–H bond in the structure –CH_2–, and the C–D bond is broken 1.6 times as readily as a C–D bond in the structure –CD_2–. Table 1 shows that the monomer yield in thermal degradation of polystyrene and poly(β-deuterostyrene) at 350 °C in vacuum is the same (42%). Monomer yield from poly(β-deuterostyrene) should have been much less than that from polystyrene if the abstraction of hydrogen from –CDH– is easier than from –CH_2–.

Table 2

Position	g, C–H	g′, C–D	Isotope effect[a]
α	9.9	3.5	2.8
β	8.7	3.25	2.7

[a] Isotope effect = g(C–H)/g′(C–D)

If a maximum in the rate versus conversion curve is an indication or measure of probability of transfer due to the presence of abstractable hydrogen atoms, then the greatest maximum should correspond to the highest degree of hydrogen transfer which in turn suggests the presence of largest number of abstractable hydrogen atoms in the polymer. Consequently, the rate versus conversion curve for $(CFHCH_2)_n$ should have the greatest maximum in the series: $(CF_2CF_2)_n$, $(CF_2CFH)_n$, $(CF_2CH_2)_n$, and $(CFHCH_2)_n$. Contrary to this, the curve for $(CFHCH_2)_n$ has the smallest maximum[64]. However, transfer seems to be very important and predominant when a reaction is carried out under high pressure. Thus, branching in polyethylene is believed to be the result of the high pressure under which the polymerization reaction is carried out. Under high pressure, the radicals remain too close to each other and the probability of radical-radical and/or radical-molecule interaction becomes greater. As a result, the transfer and termination reactions become predominant. On the other hand, thermal degradation reactions are carried out in vacuum or in an inert atmosphere. Also, the products of degradation are continuously withdrawn from the reaction zone for analysis. Under these conditions, the probability of interaction between monomer radicals or between monomer radicals and other substances would be very small, unless the monomer or the radical is trapped into the solid or the molten polymer. However, the studies of degradation of poly(α-methylstyrene) in solution in various solvents by Grant et al.[33] have shown that the monomer yield is the same as that obtained by Brown and Wall[17] using bulk molten polymer and that the solvent transfer is unimportant in the solvent systems used, because the rate of depolymerization is not a function of the solvent used.

Replacing the C–H bond in the α-position of styrene and methyl acrylate with the C–CH_3 bond raises the yield of monomer to about 100% in both cases. This is attributed to the absence of most easily abstractable tertiary hydrogen. In contrast, monomer yield in thermal degradation of polyisobutylene, which contains no tertiary hydrogen, is only 18–32%. Finally, the zip lengths are calculated from the kinetic data containing rate constant for first-order reaction and the activation energy both of which are unreliable quantities. The unreliability of these two quantities has been discussed under activation energy below.

B. Activation Energy

Attempts have been made to link the monomer yield with the activation energy involved in the overall degradation reaction. Grassie and Melville[87] have observed a marked decrease in the rate of production of monomer in thermal degradation of poly(methyl methacrylate) in vacuum as the reaction proceeds. This has been attributed to the increase in the energy of activation in the course of the reaction. The energy of activation of the reaction in its initial stages has been found to be 31.00 ± 2.00, while Jellinek and Luh[88] have found the values 68–62 kcal for the degradation of the same polymer at 300–375 °C in absence of air. The overall energy of activation for the isotactic polymer has been found to be the same as that of the syndiotactic polymer. These results are obtained on the assumption that the various energies of activation are the same for either type of polymer. The value of the expression $E_d - (E_t/2)$ have been found to be 35.92 kcal/mol which is different from the value (8 kcal) obtained for the same expression previously[89]. The difference between the two values has been attributed to different conditions under which the two sets of experiments have been conducted. The different melt viscosities of the various polymers may also influence the rate constants. The slowness of the reaction of atactic polymer also influences the energies of activation. The energies of activation are similar to that of the stereospecific polymers, that is, random initiation, depropagation, and disproportionation. At 60% conversion, the chain length of the first atactic fraction is probably still larger than $\bar{\varepsilon}$, whereas in the second case, $\bar{\varepsilon} > \bar{P}_t$. If this is so, then $(E_d - E/2) = (E_a + E/2) = 34.7$ kcal/mol, which is smaller to the result obtained for the stereospecific polymers but again does not agree at all with the value 8 kcal/mol. This shows that there are discrepancies in the reported values of the activation energies for the degradation of poly(methyl methacrylate).

Another example is the degradation of polyethylene where various values for the activation energy have been reported. Oakes and Richards[63)] have obtained values between 60 and 70 kcal from the initial molecular weight drop, but the values have been found to be independent of initial molecular weight. Jellinek[90)] obtained 45 and 66 kcal from initial stages of volatilization, but the values have been found to be dependent on molecular weight of the sample. Madorsky[68)] has obtained 45 kcal from the extrapolation of the initial part of the rate curve to zero conversion and 68 kcal from the later stages of the reaction, while Wall et al.[64)] obtained 68–72 kcal. These discrepancies may be attributed to flaws inherent in kinetic data. Kinetic studies on thermal degradation are based on the assumption that in an assembly of polymer molecules with a first-order rate constant, k, for scission, one can write $\alpha = 1 - e^{-kt}$ for the disappearance of bonds. Here α is the fraction of total initial bonds ruptured. That is, expressions for rate constant have been deduced on the assumption that the degradation reactions are of first order and, as such, the Arrhenius equation for activation energy is applicable. But the experimental results seem to contradict this. Hall[41)] has observed in the degradation of polystyrene at 340 °C in vacuum that the rate of monomer production follows a zero-order reaction over a wide range of the decomposition. Very low molecular-weight polymers have shown an entirely different kinetic behavior. The reaction is first-order up to about 30% degradation after which, as in the degradation of poly(methyl methacrylate), the formation of monomer decreases sharply. According to Jellinek[40)], the thermal degradation of polystyrene in vacuum at 250°–340 °C is of zero-order up to 60–80% monomer production. The rate constants for the same weight of fractions of different chain-length are inversely proportional to the chain-length. Recent studies[88, 91)] on thermal degradation of polystyrene have revealed that the degradation occurs by two reaction mechanisms. At lower temperatures the reaction is zero-order with an average activation energy of about 33 kcal, while at higher temperatures it is first-order with an average activation energy of about 50 kcal. Kokta et al.[91)] have found the dependence of the activation energy on molecular weight in both stages, while Funt and Magill[92)] have observed no dependence of activation energy on molecular weight. Degradation of polyethylene over the temperature range of 375°–435 °C in vacuum has been found to be zero-order but the activation energy increases with increasing molecular weight[40)], whereas high pre-exponential factor of 10^{16} has been found in the degradation of linear polyethylene[64)]. The degradation of poly(methyl methacrylate) is suspected to be a first or higher-order reaction[40)], while Grassie and Melville[15)] have presumed the initial fission reaction to be unimolecular, but their calculated value for the normal factor for unimolecular reaction is considerably less than 10^{13}. Cowley and Melville[16)] have found a very low pre-exponential factor for the rate constant with poly(methyl methacrylate). Jellinek and Luh[88)] have found an unusually high value for the pre-exponential factor, such as, of the order of 10^{21}, for the rate constant of the degradation of the same polymer. They have also observed that the first-order plots for the experiments at 300° and 375 °C show some curvature, indicating that the reactions are not strictly first-order. The degradation of poly(α-methylstyrene) over a temperature range of 280°–363 °C has been found to be zero-order up to 30% monomer production[40)], but Brown and Wall[17)] have found the pre-exponential factors of the rate constants for the same polymer quite high, of the order of 10^{18}. The presence of intermediate products in thermal degradation of the vinyl polymers also makes the kinetics complicated[73)]. It, therefore, seems that the degradation reactions are more complicated than thought of. However, some improvement could probably be made by taking into consideration the resonance, polar and steric effects, the first and the second parameters may influence the activation energy, while the third one is felt more in the frequency factor of the rate constant[94)].

C. Intramolecular Cyclization

It is evident that branches in polyethylene are produced by an intramolecular hydrogen transfer involving a transient cyclic state[95, 96)]. In vacuum degradation of a polymer, the formation of a cyclic transition state is possible only if the free radical generated is in the immediate vicinity of the polymer chain. One of the parameters which may determine the closeness of the radical to the polymer chain is the crystal structure of the polymer. The polyolefins take helical conformation in the crystalline phase, since this arrangement allows closer packing without appreciable distortion of chain bonds. If the side group is not too bulky, such as in isotactic polypropylene, polystyrene, polyethylene, etc., the helix has exactly three units per turn. With more bulky side groups such as in isotactic poly(methyl methacrylate) and polyisobutylene, there are five units in two turns and eight units in five turns, respectively. However, at temperatures of degradation reactions, the rotational barrier may be overcome and the syndiotactic and isotactic placements may be equally favorable. Actually, this helical coiling shape is reported to persist through the molten state up to the gaseous state[97–100)]. The tendency of flexible molecules to coil back on themselves has been observed in the reactions of carbenes[101)], in pyrolysis of carbon compounds[102)], and in the production of alicyclic[103)] and aromatic[104)] compounds.

This type of coiled conformation, in which a certain number of monomer units are present in a certain number of turns, would cause the terminal carbon atom of a free radical, formed by the scission of a carbon-carbon bond in the polymer chain backbone, to be in close proximity to, and to interact with, a specific carbon atom, or a hydrogen atom linked to a specific carbon atom, in the turn. Thus, in an isotactic polyethylene molecule, which contains three monomer units per turn and is represented by the structure in Fig. 2, the scission of the $C_{\#6} - C_{\#7}$ bond would bring $C_{\#6}$ in close proximity to $C_{\#1}$ or $C_{\#2}$ or the hydrogen atoms linked to any one of them. If $C_{\#6}$ containing the unpaired electron attacks the $C_{\#1}$, a six-membered ring may be formed [eq. (15)]. It has been observed that in an intramolecular cyclization, the fastest reactions are those which proceed via six-membered rings[105)]. However, cyclohexane (I) may lose one or more hydrogen atoms to a free radical and form hexene, hexadiene, hexane, and other compounds as shown below.

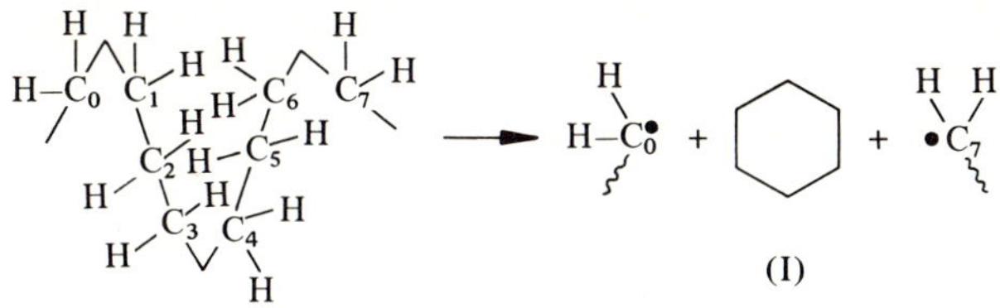

Fig. 2 (I) (15)

(I) 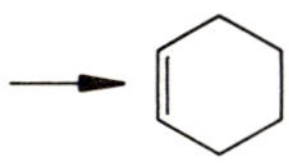 + $2H^\bullet$ (16)

(II)

(II) → $CH_2{=}CH{-}CH{=}CH{-}CH_2{-}CH_3$ (17)

(II) $\xrightarrow{2H^\bullet}$ $CH_3{-}CH{=}CH{-}CH_2{-}CH_2{-}CH_3$ (18)

(II) → $CH_2{=}CH{-}CH_2{-}CH_2{-}CH_2{-}CH_3$ (19)

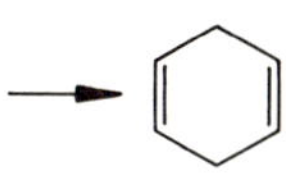 + 2H (20)

(III)

(III) → $\dot{C}H_2{-}CH{=}CH{-}CH{=}CH{-}\dot{C}H_2$ (21)

(IV)

(IV) $\xrightarrow{2H}$ $CH_3{-}CH{=}CH{-}CH{=}CH{-}CH_3$ (22)

(II) → $CH_2{=}CH{-}CH{=}CH_2$ + $CH_2{=}CH_2$ (23)

(II) → $CH_3{-}CH{=}CH{-}CH_3 + CH_3{-}CH_3$ (24)

(II) → $CH_2{=}CH{-}CH_3 + CH_3{-}CH_2{-}CH_3$ (25)

If the $C_{\#6}$ attacks the $C_{\#2}$, a five-membered ring (V) is formed, which may undergo degradation to aliphatic hydrocarbons as follows.

 + 2H (26)

(V) (VI)

$$(VI) \longrightarrow \dot{C}H_2{-}CH_2{-}CH{=}CH{-}\dot{C}H_2 \quad (27)$$

(VII)

$$(VII) \xrightarrow{2H} CH_3{-}CH_2{-}CH{=}CH{-}CH_3 \quad (28)$$

$$\xrightarrow{2H} CH_3{-}CH_2{-}CH_2{-}CH_2{-}CH_3 \quad (29)$$

$$(VII) \longrightarrow CH_2{=}CH{-}CH_2{-}CH{=}CH_2 \quad (30)$$

Hydrogen abstraction by the $C_{\#6}$ from the $C_{\#1}$ or $C_{\#2}$ may also lead to the formation of these products. All of the aliphatic hydrocarbons formed according to eqs. (19) through (32) have been reported[64] to be the volatile products of thermal degradation of polyethylene.

As isotactic polystyrene has a similar crystal structure to that of polyethylene, the concept of the intramolecular cyclization may be applied to the degradation of the former to account for the formation of the products of degradation. Let the three-unit one turn radical form the cyclic compound (VIII) by the interaction of the $C_{\#1}$ and $C_{\#6}$ atoms. The cyclic compound (VIII) may lose hydrogens to other free radicals to form 1,3,5-triphenylbenzene (IX).

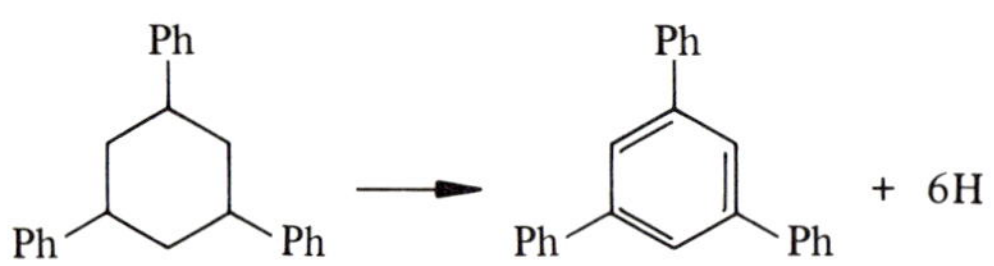

(31)

(VIII) (IX)

If the terminal carbon atom abstracts a hydrogen from the C_0 or $C_{\#1}$ atom, the degradation may follow the course given below.

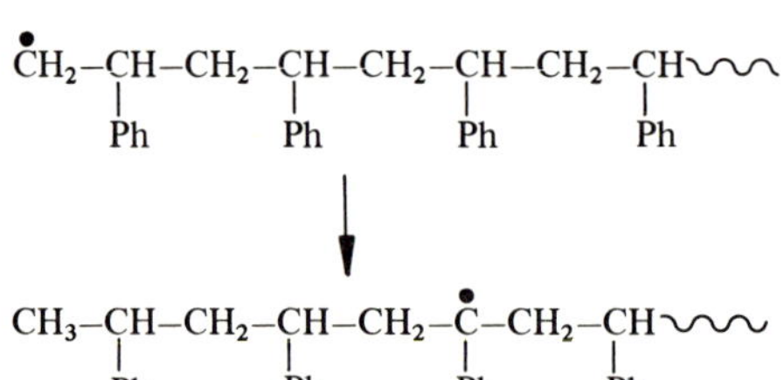

(32)

(X)

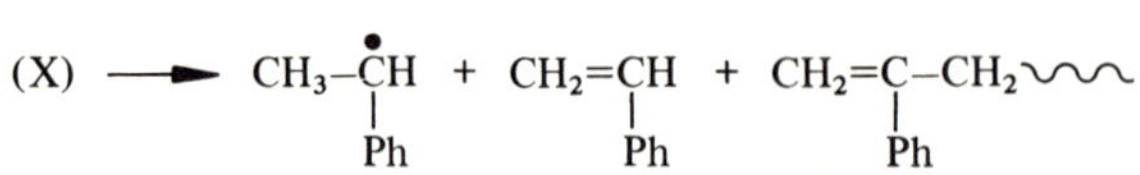

(33)

(XI)

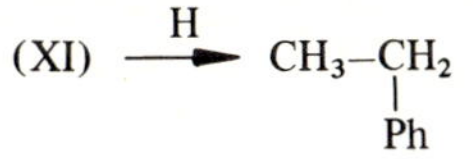
(34)

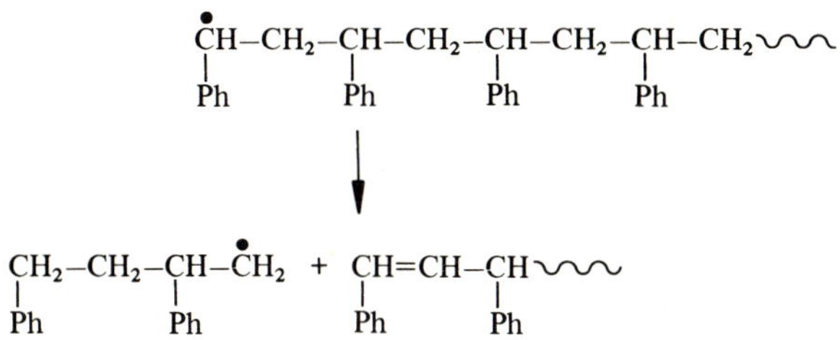
(35)

(XII)

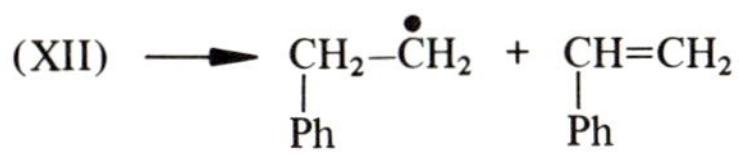
(36)

(XIII)

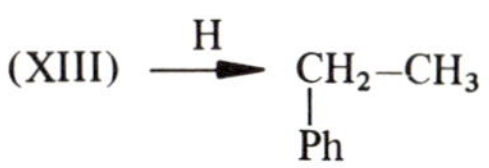
(37)

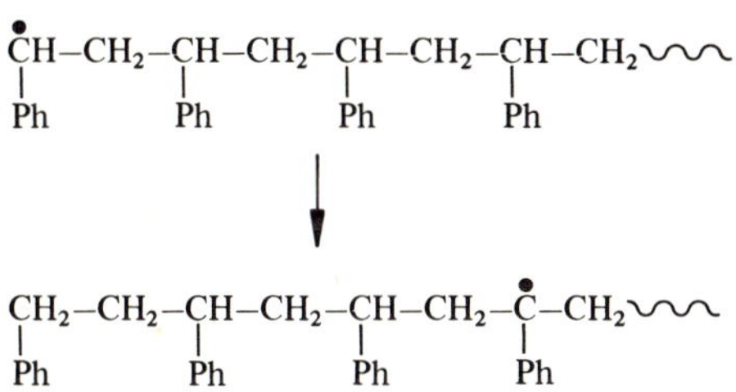
(38)

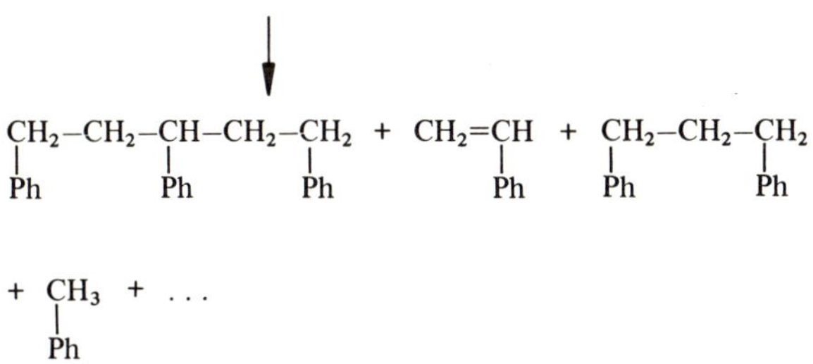
(39)

All of the products 1,3,5-triphenylbenzene, 1,3-diphenylpropane, 1,3,5-triphenylpentane, ethylbenzene, methylbenzene, and styrene are volatile products of thermal degradation of polystyrene[42]. This kind of intramolecular cyclization is not likely to be important in thermal degradation of poly(methyl methacrylate) and poly(α-methylstyrene), because the reactive site is linked to two bulky substituents which restrict the rotation of the terminal carbon atom containing the unpaired electron and thus prevent it from coming into proper orientation to attack the other carbon atoms in the chain.

D. Steric Effects

According to Jellinek[69], three parameters may determine mainly the degradation characteristics of a polymer. These are:
(a) the size of the substituents X and Y in $-C(X)(Y)-CH_2-$,
(b) presence or absence of tertiary hydrogen atoms in the polymer chain backbone, and
(c) the resonance energy of radicals formed during degradation.
If X and Y are large, then steric hindrance would be prevalent in the polymer chain leading to strain and weakening of carbon-carbon bonds. It then follows that a polymer containing α-methyl groups would produce, during thermal degradation, larger quantities of monomer. But polyisobutylene containing α-methyl group in the polymer chain backbone produces up to 32 weight % of monomer, while due to the presence of α-methyl group, a much higher monomer yield is expected.

Given that two radicals or a radical and a molecule are in close proximity to each other and that the necessary activation energy is provided, the reaction between them would be controlled by the extent of steric hindrance at the reactive site. If no bulky substituents are present at the active site, the radicals would react with each other very fast. For example, ethylene diradicals, $\dot{C}H_2{-}\dot{C}H_2$, having only small hydrogen atoms at the reactive sites would react among themselves or with other species very fast. Thus, two ethylene diradicals, evolved during degradation of polyethylene, may react with each other forming ultimately butane or butylene as follows.

$$\dot{C}H_2{-}\dot{C}H_2 + \dot{C}H_2{-}\dot{C}H_2 \rightarrow \underset{\text{(XIV)}}{\dot{C}H_2{-}CH_2{-}CH_2{-}\dot{C}H_2} \quad (40)$$

$$\text{(XIV)} \rightarrow CH_3{-}CH_2{-}CH{=}CH_2 \quad (41)$$

$$\text{(XIV)} \xrightarrow{2\,H} CH_3{-}CH_2{-}CH_2{-}CH_3 \quad (42)$$

$$\dot{C}H_2{-}\dot{C}H_2 + 2\,H \rightarrow CH_3{-}CH_3 \quad (43)$$

On the other hand, if there are bulky substituents at the reactive sites, the proper configuration and orientation towards each other of the reactive sites may take a long time. Even if the two reactive sites acquire suitable orientation towards each other, the reaction may be delayed or stopped completely by the steric hindrance. Thus, styrene monomer unit has an aromatic ring and a hydrogen atom at the reactive site. The hydrogen atom offers very little steric hindrance but the aromatic ring offers a good deal of steric hindrance. Therefore, the rate of the reaction between a styrene radical and another radical or molecule would be smaller than that between two ethylene radicals. If the two hydrogen atoms at the reactive site of an ethylene radical are replaced by one phenyl group and one methyl group, then the rate of the reaction would further be decreased by the greater steric hindrance of the two bulky substituents. Therefore, the yield of monomer would increase from polyethylene to polystyrene to poly(α-methylstyrene), from less than 1% monomer yield to about 100% monomer yield in poly(α-methylstyrene). Both poly(α-methylstyrene) and poly(methyl methacrylate) contain bulky substituents. It has been stated above for intramolecular cyclization that in the formation of a six-membered ring it is difficult for $C_{\#6}$ to rotate and attack $C_{\#1}$ or $C_{\#2}$ atom or hydrogens attached to them if too many bulky groups are linked with $C_{\#6}$. This argument explains, at least partly, the formation of very little non-monomeric products in thermal degradation of poly(α-methylstyrene) and poly(methyl methacrylate).

E. Resonance

According to Jellinek[69)], the energy required to break a bond would be much smaller than is normally the case if a radical formed during degradation of a polymer can resonate between a number of different structures. In the case of the allyl radical $H_2C{=}CH{-}CH_2{-}$, the resonance stabilization energy is 19 kcal. To break an ordinary carbon-carbon single bond in a polymer chain backbone $-CH_2{-}CH_2{-}CH_2{-}$, an amount of energy equal to 81 kcal is required. However, to break a carbon-carbon single bond in the β-position to a double bond an amount of energy equal to 81–19 = 62 kcal would be needed. This explains the ease with which a carbon-carbon single bond in the polymer chain backbone is cleaved, but the role of the resonance stabilization energy in determining the formation of monomer has not been discussed.

It has been revealed by X-ray studies[106, 107)] that in polyethylene the carbon-carbon bond distance is 1.54 Å and the valence angle 108°, suggesting that the carbon atom in the polymer chain is tetrahedral. As all of the polyolefins including polyethylene possess helical coiled structure, the tetrahedral stereochemistry for the carbon atom in the polymer chain would persist in all of them. With bulky pendant groups linked to the polymer chain, this tetrahedral structure is most likely to be under considerable strain. For example, polyethylene, through a carbon-carbon bond scission in the polymer chain, forms the radical (XV) in which the unpaired electron and two C–H bonds on $C_{\#1}$ are in trans or gauche conformation with the two $C_{\#2}$–H bonds and the $C_{\#2}$–$C_{\#3}$ bond. The radical end

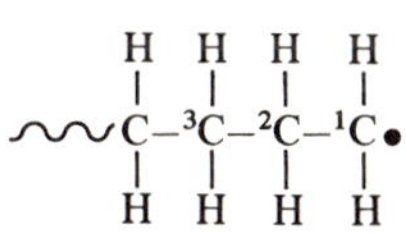

(XV)

can also rotate freely about the $C_{\#1}$–$C_{\#2}$ bond. As there are infinite number of rotational conformations of the terminal $-\dot{C}H_2$ group about the $C_{\#1}$–$C_{\#2}$ bond, there is only one in this infinite number of probabilities where the lone electron would come into proper orientation to form a π-bond between $C_{\#1}$ and $C_{\#2}$. Even if the lone electron comes into proper orientation, the formation of the ethylene molecule would be discouraged due to its very small resonance stabilization energy. Resonance stabilization energies, from the calculated[1] and observed[2] heats of hydrogenation, for ethylene, propylene, and isobutylene are given in Table 3.

Thus, as the hydrogen in ethylene is replaced stepwise with a methyl radical, the resonance energy increases. This is attributed to the increase in the number of canonical

1 Calculated from bond energy
2 Ref. 108

Table 3

Compound	Heat of hydrogenation (kcal) Calculated	Observed	Resonance Energy (kcal)
Ethylene	31.75	32.80	−1.05
Propylene	31.75	30.10	+1.65
Isobutylene	31.75	28.40	+3.35

structures from ethylene to isobutylene. The methyl group, through hyperconjugation, delocalizes the electron and, therefore, the number of canonical forms increases with the increase in the number of methyl groups. Consequently, thermodynamic stability increases from ethylene to isobutylene and with this the monomer yield increases from polyethylene to polyisobutylene. Replacing one hydrogen in ethylene with a six-membered aromatic ring, we obtain styrene which has more canonical forms than isobutylene and is, therefore, thermodynamically more stable than the latter. If the α-hydrogen in styrene is replaced with a methyl group, the resonance stabilization of the resulting α-methylstyrene becomes greater than that of styrene through the participation of the methyl group in resonance conjugation with the aromatic ring and the ethylenic double bond. Consequently, the monomer yield increases in the order:

ethylene $<$ propylene $<$ isobutylene $<$ styrene $<$ α-methylstyrene.

However, there is another parameter which seems to prepare the ground for resonance effects to occur. The bulky groups linked to the terminal $C_{\#1}$ atom of the free radical (XVI) formed by scission of a carbon-carbon bond in the polymer chain backbone of poly(α-methylstyrene), can have more room and hence, can lessen steric interaction among themselves as well as with the other groups and atoms in the polymer chain when

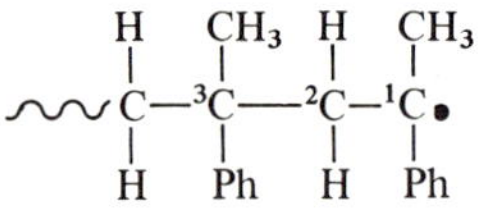

(XVI)

they are coplanar at angles of 120° to each other than when they are at the corners of a tetrahedron and are 109° apart. When the system $C_{\#2}$–$\dot{C}_{\#1}$ becomes coplanar, the p-orbital on the $C_{\#1}$ atom in which the unpaired electron resides becomes perpendicular to this planar structure and becomes parallel to one of the p-orbitals on the $C_{\#2}$ atom, thus enabling delocalization of the lone electron and the system $C_{\#2}$=$C_{\#1}$ becomes stabilized. The $C_{\#3}$–$C_{\#2}$ bond undergoes scission and a molecule of α-methylstyrene is evolved. This process continues till the whole of the chain radical (XVI) has completely unzipped into monomer. Similar explanations may be advanced for 100% yield of monomer in thermal degradation of poly(methyl methacrylate). Recently, Sohma and Sakaguchi[109] have found the radical (XVII) to be more reactive to oxygen than (XVIII).

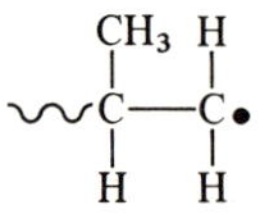

(XVII)

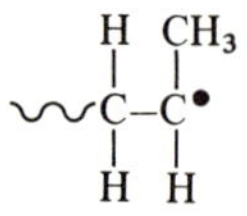

(XVIII)

The authors have offered two plausible explanations for this difference in reactivity:
(a) diminished accessibility of an oxygen molecule to the site of the unpaired electron of the radical (XVIII) and
(b) a more stabilized structure of the radical (XVIII) because of the hyperconjugation of the p_π orbital of the unpaired electron to the methyl group.

III. Conclusion

The studies of the kinetics of decomposition of polymers have become instrumental in elucidating the mechanism of degradation. Such studies are based on a general chain mechanism consisting of initiation, depropagation, transfer, and termination reactions. It seems that none of these theories has been able to advance the subject of thermal degradation of high polymers for lack of satisfactory experimental data. There is large difference in monomer yields on thermal degradation of various polymers under identical conditions and for a certain polymer the experimental results are similar in various studies. But no completely satisfactory interpretations have been produced, capable of explaining all the features of the degradation, some of which appear to be contradictory.

It may, therefore, be concluded that such parameters as intramolecular cyclization, intramolecular hydrogen abstraction, steric effects, and resonance effect can most satisfactorily explain the yields of monomers in thermal degradation of polyolefins. Unfortunately, except for hydrogen abstraction, three other parameters have not been taken into consideration in deducing expressions of rate constants for thermal degradation of polyolefins. The resonance effect should influence the activation energy of the degradation reaction and the steric effects are more prominent in the frequency factors of the rate constants[94].

Acknowledgement. The author is indebted to the referee for very useful suggestions.

IV. References

1. Williams, C. G.: Phil. Trans. *150,* 241 (1860)
2. Meyer, K. H., Hopff, H., Mark, H.: Ber. dhch. chem. Ges. *62,* 1103 (1922)
3. Kuhn, W.: ibid. *63,* 1503 (1930)
4. Kuhn, W., Freudenberg, K.: ibid. *63,* 1510 (1930)
5. Simha, R.: J. Appl. Phys. *12,* 569 (1941)
6. Montroll, E., Simha, R.: J. Chem. Phys. *8,* 721 (1940)
7. Schulz, G. V.: Z. Physik. Chem. *B 52,* 50 (1942)
8. Votinov, A., Kobeko, P., Marei, F.: J. Phys. Chem. U.S.S.R. *16,* 106 (1942)
9. Lehmann, F. A., Brauer, G. M.: Anal. Chem. *33,* 673 (1961)
10. Madorsky, S. L.: Thermal Degradation of Polymers, New York: Wiley-Interscience 1964
11. Shimizu, Y., Munson, B.: J. Polym. Sci., Polym. Chem. Ed. *17,* 1991 (1979)
12. McNeill, I. C. et al.: J. Polym. Sci., Polym. Chem. Ed. *15,* 2381 (1977)
13. Grassie, N., Melville, H. W.: Proc. Roy. Soc. (London) *A 199,* 14 (1949)
14. Grassie, N., Melville, H. W.: Proc. Roy. Soc. (London) *A 199,* 1 (1949)

15. Grassie, N., Melville, H. W.: Proc. Roy. Soc. (London) *A 199,* 24 (1949)
16. Cowley, P. R. E. J., Melville, H. W.: Proc. Roy. Soc. (London) *A 210,* 461 (1951)
17. Brown, D. W., Wall, L. A.: J. Phys. Chem. *62,* 848 (1958)
18. Bywater, S., Black, P. E.: J. Phys. Chem. *69,* 2967 (1965)
19. MacCallum, J. R.: Trans. Faraday Soc. *59,* 2099 (1963)
20. MacCallum, J. R.: Makromol. Chem. *83,* 137 (1965)
21. Barlow, A. et al.: Polymer *8,* 537 (1967)
22. Jellinek, H. H. G., Luh, M. D.: Polymer Reprints, Vol. 8, p. 533, Amer. Chem. Soc. Meeting, Miami Beach, April, 1967
23. Bagby, G., Lehrle, R. S., Robb, J. C.: Makromol. Chem. *119,* 122 (1968)
24. Reich, L., Stivala, S. S.: Elements of Polymer Degradation, p. 168, London: McGraw-Hill 1971
25. McNeill, I. C.: Eur. Polym. J. *4,* 21 (1968)
26. Bamford, C. H., Tipper, C. F. H.: Comprehensive Chemical Kinetics, p. 53, Vol. 14, New York: Elsevier Scientific Publ. Comp. 1975
27. Cameron, G. G., Kerr, G. P.: Makromol. Chem. *115,* 268 (1968)
28. Davis, L. A.: J. Polym. Sci., Polym. Phys. Ed. *12,* 75 (1974)
29. Simha, R., Wall, L. A., Blatz, P. J.: J. Polymer Sci. *5,* 615 (1950)
30. Simha, R., Wall, L. A.: J. Polymer, Sci. *6,* 39 (1951)
31. Simha, R., Wall, L. A.: J. Phys. Chem. *56,* 707 (1952)
32. Bywater, S., Black, P. E.: J. Phys. Chem. *69,* 2967 (1965)
33. Grant, D. H., Vance, E., Bywater, S.: Trans. Faraday Soc. *56,* 1697 (1960)
34. Cowie, J. M. G., Bywater, S.: J. Polymer Sci. *54,* 221 (1961)
35. Jellinek, H. H. G., Luh, M. D.: Eur. Polym. J., Suppl. 149 (1969)
36. Madorsky, S. L.: J. Res. Natl. Bur. Stand. *62,* 219 (1959)
37. Staudinger, H., Steinhofer, A.: Liebigs Ann. Chem. *517,* 35 (1935)
38. Jellinek, H. H. G.: Trans. Faraday Soc. *40,* 266 (1944)
39. Jellinek, H. H. G.: Trans. Faraday Sci. *44,* 345 (1948)
40. Jellinek, H. H. G.: Disc. Faraday Soc. *2*, 397 (1947)
41. Hall, R. W.: Disc. Faraday Soc. *2,* 396 (1947)
42. Bolker, H. I.: Natural and Synthetic Polymers, An Introduction, p. 181, New York: Marcel Dekker 1974
43. Grassie, N., Kerr, W. W.: Trans. Faraday Soc. *55,* 1050 (1959); ibid. *53,* 234 (1957)
44. Gordon, M.: ibid. *53,* 1462 (1957)
45. Wall, L. A., Brown, D. W., Hart, V. E.: J. Polymer Sci. *15,* 157 (1955)
46. Wall, L. A.: in G. M. Kline (Ed.), Analytical Chemistry of Polymers, Part II, New York: Wiley 1962
47. Cameron, G. G., Grassie, N.: Polymer *2,* 367 (1961)
48. Jellinek, H. H. G., Spencer, L. B.: J. Polymer Sci. *8,* 573 (1952)
49. Madorsky, S. L. et al.: J. Res. Natl. Bur. Stand., Sect. A, *66,* 307 (1962)
50. Cameron, G. G.: Makromol. Chem. *100,* 255 (1967)
51. Cameron, G. G., McWalter, I. T.: Eur. Polym. J. *6,* 1601 (1970)
52. Cameron, G. G., McWalter, I. T.: IUPAC Internat. Symp. Macromol. Chem., Budapest, 1969, Vol. V, Akademiai Kiado, Budapest, 1969, p. 311
53. Cameron, G. G., Kerr, G. P.: Eur. Polym. J. *4,* 709 (1968)
54. Cameron, G. G., Kerr, G. P.: Eur. Polym. J. *6,* 423 (1970)
55. Boon, J., Challa, G.: Makromol. Chem. *84,* 25 (1965)
56. Nakajima, A., Hamada, F., Shimizu, T.: Makromol. Chem. *90,* 229 (1966)
57. Madorsky, S. L.: Thermal Degradation of Organic Polymers, p. 100, New York: Interscience 1964
58. Wall, L. A., Straus, S.: J. Polymer Sci. *44,* 313 (1960)
59. Igarashi, S., Kambe, H.: Bull. Chem. Soc. Jap. *37,* 176 (1964)
60. Moiseev, V. D.: Sov. Plast. 6 (1964)
61. Khloplyaukina, M. S., Neiman, M. B., Moiseev, V. D.: Sov. Plast. 11 (1961)
62. Tsuchiya, Y., Sumi, K.: J. Polymer Sci., Part A-1, *6,* 415 (1968)
63. Oakes, W. G., Richards, R. B.: J. Chem. Soc. (London) 2929 (1949)
64. Wall, L. A. et al.: J. Amer. Chem. Soc. *76,* 3430 (1954)

65. See Ref. 26
66. Wall, L. A., Florin, R. E.: J. Res. Natl. Bur. Stand. *60,* 451 (1958)
67. Madorsky, S. L. et al.: J. Res. Natl. Bur. Stand. *42,* 499 (1949)
68. Madorsky, S. L.: J. Polymer Sci. *9,* 133 (1952)
69. Jellinek, H. H. G.: Degradation of Vinyl Polymers, p. 161, New York: Academic Press 1955
70. Holmstrom, A., Sorvik, E. M.: J. Polymer Sci., Polymer Symposia No. 57, 33 (1976)
71. Rideal, G. R., Padget, J. C.: J. Polymer Sci., Polymer Symposia No. 57, 1 (1976)
72. Seeger, M., Barrall, E. M.: J. Polym. Sci., Polym. Chem. Ed. *13,* 1515 (1975)
73. Seeger, M., Gritter, R. J.: J. Polym. Sci., Polym. Chem. Ed. *15,* 1393 (1977)
74. Michajlov, L., Zegenmaier, P., Cantow, H.-J.: Polymer *9,* 325 (1968)
75. Madorsky, S. L. et al.: J. Polymer Sci. *4,* 639 (1949)
76. Van Schooten, J., Evenhuis, J. K.: Polymer *6,* 343 (1965)
77. Seeger, M., Cantow, H.-J.: Makromol. Chem. *176,* 1411 (1975)
78. Seeger, M., Exner, J., Cantow, H.-J.: Paper Presented at Internat. Symp. Macromolecules (IUPAC), Helsinki, 1972; Reprints *5,* 177
79. Lenz, R. W.: Organic Chemistry of Synthetic High Polymers, p. 742, New York: Interscience 1967
80. Grassie, N.: Chemical Reactions of Polymers (E. M. Fettes, Ed.) p. 565 ff, New York: Interscience 1964
81. Mita, I.: in Aspects of Degradation and Stabilization of Polymers (H. H. G. Jellinek, Ed.) Chapter 6, p. 261, New York: Elsevier Scientific Publ. Comp. 1978
82. Wall, L. A.: SPE J. *16(8),* 1 (1960)
83. Ref. 79, Chapter 18, p. 741
84. Ref. 81, p. 263
85. Madorsky, S. L., Straus, S.: J. Res. Natl. Bur. Stand. *40,* 417 (1948)
86. Wall, L. A., Brown, D. W.: J. Phys. Chem. *61,* 129 (1957)
87. Grassie, N., Melville, H. W.: Disc. Faraday Soc. *2,* 378 (1947)
88. Jellinek, H. H. G., Luh, M. D.: J. Phys. Chem. *70,* 3672 (1966)
89. Jellinek, H. H. G.: J. Polymer Sci. Lett. *2,* 457 (1964)
90. Jellinek, H. H. G.: J. Polymer Sci. *4,* 1, 13 (1949)
91. Kokta, B. V., Valada, J. L., Martin, W. N.: J. Appl. Polym. Sci. *17,* 1 (1973)
92. Funt, J. M., Magill, J. H.: J. Polym. Sci., Polym. Phys. Ed. *12,* 217 (1974)
93. Kline, G. M.: Analytical Chemistry of Polymers. Part II. Analysis of Molecular Structure and Chemical Groups, p. 185, New York: Interscience 1962
94. Ref. 79, p. 279
95. Raedel, M. J.: J. Amer. Chem. Soc. *75,* 6110 (1953)
96. Willbourn, A. H.: J. Polymer Sci. *34,* 369 (1959)
97. Meyerson, S., Leitch, L. C.: J. Amer. Chem. Soc. *93,* 2244 (1971)
98. Dinh-Nguyen, N. et al.: Ark. Kemi *18,* 393 (1961)
99. Tanford, C.: Physical Chemistry of Macromolecules, Chapters 2 and 3 and references therein. New York: Wiley 1961
100. Ubbelohde, A. R., McCoubrey, J. C.: Disc. Faraday Soc. *10,* 94 (1951)
101. Moss, R. A.: Chem. Eng. News, 50 (June 30, 1969)
102. Hurd, C. D.: The Pyrolysis of Carbon Compounds. Chemical Catalog Co., New York, 1929
103. Weitkamp, A. W. et al.: Ind. Eng. Chem. *45,* 343 (1953)
104. Cady, W. E., Launer, P. J., Weitkamp, A. W.: Ind. Eng. Chem. *45,* 350 (1953)
105. Bell, R. P., Fluendy, M. A. D.: Trans. Faraday Soc. *59,* 1623 (1963)
106. Bunn, C. W.: ibid. *35,* 482 (1939)
107. Brill, R.: Z. Physik. Chem. *B53,* 66 (1943)
108. Morrison, R. T., Boyd, R. N.: Organic Chemistry, Third Ed., p. 183, Boston: Allyn and Bacon, Inc. 1973
109. Sohma, J., Sakaguchi, M.: Adv. Polym. Sci., Vol. 20, p. 111, Heidelberg, New York: Springer-Verlag 1976

Received May 26, 1982
Charles G. Overberger (editor)

Entangled Linear, Branched and Network Polymer Systems – Molecular Theories

William W. Graessley*

Cavendish Laboratory, Madingley Road, Cambridge CB 3 OHE, England

This article deals with recent work on the theory of entanglement effects in polymer rheology, in particular, the reptation idea of deGennes and the tube model of Doi and Edwards. Predictions which depend on the independent alignment approximation are omitted. Attention is focussed primarily on linear viscoelastic properties, macromolecular diffusion, relaxation following step strains and network properties. Theoretical predictions and experimental observations are discussed for monodisperse linear and star-branched polymer liquids. Effects due to chain length distribution, relaxation behavior of unattached chains in networks, and the equilibrium elasticity of networks are also considered. The results suggest a need to consider relaxation mechanisms in addition to simple reptation as well as certain modifications in the tube model itself. The effect of fluctuations in chain density along the tube is probably quite important in branched chain liquids. Considerations about lifetimes of the tube defining constraints seem necessary to account for polydispersity effects and for differences between relaxation rate of chains in liquids and in networks. A modified tube model is proposed which gives a somewhat better description of elasticity in entangled networks while still producing the Doi-Edwards expression for stress in entangled liquids. Experimental results so far are qualitatively consistent with the picture which is presented here. Much work needs to be done however to test for quantitative consistency.

* Permanent address: Chemical Engineering Department, Northwestern University, Evanston, Illinois 60201, U.S.A.

Advances in Polymer Science 47

List of Symbols

In general the terminology follows that of Ferry[12] for viscoelastic functions and molecular variables and Doi and Edwards[2,3] for reptation and tube model variables.

a	primitive path step length
c	polymer concentration (mass/volume)
D	macroscopic diffusion coefficient
D*	diffusion coefficient along the primitive path
D_r	diffusion coefficient from the Rouse model
d	mesh size, length of a primitive segment (Pt. II)
E	displacement gradient tensor
F	free energy
F(t)	fraction of initial primitive path steps which are still occupied
f	branch point functionally, number of strands emanating from the same junction point
k	Boltzmann constant
L	average primitive path length
L_o	Polymer chain length
L_m	length of a primitive path with m primitive segments
M	molecular weight of the polymer
M_o	molecular weight of the monomer unit
m	number of path-occupying primitive segments
m_o	total number of primitive segments for a molecule
n_o	number of monomer units in a chain
N	mean number of primitive path steps for a molecule
N_b	mean number of primitive path steps along a branch
N_s	number of primitive path steps occupied by the matrix chains in a mixture
P_m	path length probability function
p	probability that a primitive segment selected at random is part of the surplus (loop) population
Q_m	probability that a chain selected at random contains at least m path-occupying primitive segments, $\sum_{n=m}^{mo} P_m$
$\mathbf{R}_g$	position of center-of-gravity of a molecule
R_G	universal gas constant
$\langle R^2 \rangle$, $\langle r^2 \rangle$	mean-square end-to-end distance for unperturbed chains and parts of chains

$\mathbf{r}_i$	end-to-end vector for a primitive path step
S	entropy
S_1, S_2, S_3	numerical coefficients of order unity obtains from summations
T	temperature
$\mathbf{u}$	unit tangent vector for a primitive path step
W	stored energy function
z	number of "suitably situated" constraints defining a primitive path step
z_o	total number of constraint strands defining a primitive path step
γ	strain in simple shear
ε	Curtiss-Bird link tension coefficient
ζ_o	Monomeric friction coefficient
λ	stretch ratio in uniaxial extension
λ_i	eigenvalues in the Rouse model
ν	chains per unit volume
ϱ	undiluted polymer density
σ	shear stress or tensile stress (in context)
$\sigma_{\alpha\beta}$	α, β component of the stress tensor
τ	relaxation time
τ_d	Doi-Edwards disengagement time
τ_e	Doi-Edwards equilibration time
τ_r	longest relaxation time in the Rouse model
τ_w	mean waiting time for contraint release
τ_{tr}	time characterizing the linear viscoelastic transition region
ϕ	jump frequency
ϕ	volume fraction of polymer
$\Lambda(z)$	proportionaliy constant relating τ_w and τ_d for linear chains

Introduction

The idea that entangled chains rearrange their conformations by reptation, i.e., curvilinear diffusion along their own contours, was introduced several years ago by de Gennes[1]. He considered reptation by long unattached linear chains in a medium of permanent topological obstacles, representing the strands of a crosslinked network, and showed that the macroscopic diffusion coeffient and the time for complete rearrangement of conformation would vary with chain length (or molecular weight M) as

$$D \propto M^{-2}$$

$$\tau \propto M^{3}$$

Doi and Edwards have recently developed a theory relating the dynamics of reptating chains to mechanical properties in concentrated polymer liquids[2–5]. They assumed that reptation would be the dominant motion in a medium filled with long chains even in the absence of a permanent network (2). Adapting equations from the theory of rubber elasticity they calculated the contribution of individual chains to the stress following a step strain and related the subsequent relaxation of stress to conformational rearrangement via reptation[3]. With additional assumptions, principally the "independent alignment approximation"[3], they were able to extend the analysis to arbitrary deformation histories, arriving at a constitutive equation of the BKZ type[4,5]. Subsequent work by Doi has dealt with alternative assumptions in the constitutive generalization[6], with "non-ever-increasing" strain histories where the model leads to results different from BKZ[7], and with contributions to relaxation other than simple reptation[8–10].

Doi and Edwards were able to develop an equation for the stress relaxation modulus G(t) of monodisperse entangled linear chain liquids in the terminal region without resorting to the independent alignment approximation[3]. From G(t), expressions can be obtained for the plateau modulus, the steady-state viscosity and steady-state recoverable compliance. The following dependences on chain length are obtained:

$$G_N^0 \propto M^0$$

$$\eta_0 \propto M^3$$

$$J_e^0 \propto M^0$$

The detailed equations for D, τ, G_N^0, η_0 and J_e^0 can be expressed in terms of a single reptation model parameter and the monomeric friction coefficient[11]. The latter can be estimated independently (Ref. 12, Chap. 12), and predictions of relationships among properties can be obtained by expressing the reptation model parameter in terms of one experimental property such as G_N^0.

The expression obtained in this way for the diffusion coefficient appears to work very well, providing not only the observed dependence on molecular weight but also close numerical agreement with measured values[11]. A universal expression is obtained for recoverable compliance:

$$G_N^0 J_e^0 = \frac{6}{5}$$

The constancy of this product is consistent with experiment, but the numerical factor is somewhat larger ($G_N^0 J_e^0 \approx 3$)[10]. The viscosity varies more rapidly with chain length than predicted ($\eta_0 \propto M^{3.4}$ for $M > M_c$). The predicted values (proportional to M^3) are in fact larger in the range of M covered by existing data[11].

These results for J_e^0 and η_0 may reflect the "upper limit" nature of the model. Each chain in a monodisperse liquid is assumed to rearrange its conformation by reptation alone and as though it were moving in a permanent network of obstacles. Competition from other mechanisms would of course increase the rate of relaxation and thus produce a lower viscosity. It would also increase the breadth of the relaxation spectrum and thus produce a larger value of the $G_N^0 J_e^0$ product[13].

Other evidence suggests that the reptation is only one of perhaps several mechanisms operating in non-network systems:

(i) The relaxation time of long unattached linear chains in a network is larger than in the corresponding monodisperse system (the homologous melt) by a constant factor of about 3[14,15] or even more in some cases[16].

(ii) The presence of long branches in a molecule should suppress reptation[17,18]. Nevertheless, the viscosities of star polymers (three or more long arms emanating from a single junction) remain comparable to those for linear polymers of the same molecular size even well beyond the arm lengths which would correspond to the onset of entanglement effects for linear chains[19]. On the other hand, the viscosity does eventually begin to increase very rapidly relative to linear chains as the arms become still longer[19,20]. This suggests that reptation can only begin truly to dominate relaxation even of linear polymers when the chains become rather long compared to M_c. This region of dominance by reptation is perhaps signalled indirectly by the onset of observable effects due to suppression of that motion in the case of star molecules.

(iii) It is observed that the longest relaxation times of unattached star molecules in a network are very much larger than those in the homologous melt[14], and, unlike the behavior of linear chains, the ratio of relaxation times appears to increase rapidly with arm length for star molecules. This suggests that when reptation is suppressed the connectivity of the environment begins to assume a dominating role.

(iiii) The predicted effects of molecular weight distribution in liquids are much stronger than those observed experimentally; contributions of the largest chains are weighted too heavily in the strict reptation model[11]. All molecules are assumed to relax independently and purely by reptation. In liquids this is probably unrealistic for chains which are much larger than the average.

There are in fact two additional processes for relaxation which are quite natural to consider. One is the renewal of conformation by release of constraints which confine each chain, arising from diffusion of the surrounding chains which supply the constraints[21]. This "constraint release" mechanism would operate in liquids, where the obstacles are themselves parts of reptating chains, but not in a network.

The other relaxation process is associated with fluctuations in conformation and should operate in both liquids and networks. The tube of constraints around the contour of any chain is composed of molecular strands, acting (crudely) like the bars of a cage.

Temporarily the confined chain can bulge out past the bars anywhere along the tube. Thus, some average portion of the chain will occupy the cage, and the rest will provide closed loops or surplus length protruding from the cage. The amount of surplus length will fluctuate about some average, and, correspondingly, so will the length of cage occupied by the chain. A fluctuation which temporarily shortens the occupied length will cause abandonment of parts of the cage which had been previously occupied. When that fluctuation subsides the free ends move out, but along random paths, thus creating a new conformation for those parts of the chain. This "tube leakage" process is implicit in de Gennes' original model of a reptating chain[1)], and in fact he used a simplified version of it later to discuss conformational relaxation in star molecules[17)]. Doi has also recently considered the effect of fluctuations in the occupied length on the relaxation of star molecules[10)]. Doi and Edwards omitted fluctuations and considered only simple reptation, i.e., disengagement by the average diffusive motion of each chain along its contour.

This paper presents a model for entangled systems which incorporates all these features. The object is to provide a single framework for analysis of data on diffusion, linear viscoelastic behavior, finite single-step-strain response and network elasticity in a wide variety of structures, i.e., linear, star-branched, network and mixed length systems. The independent alignment approximation is not used. The first part is a brief recapitulation of the de Gennes and Doi-Edwards ideas. The second part presents the proposed model and considers behavior in topologically invariant surroundings: equilibrium network elasticity, and the effects of tube leakage (occupied length fluctuations) on diffusion and stress relaxation for unattached linear and star-branched molecules in networks. The third part considers the effects of constraint release, i.e., the additional contributions to diffusion and relaxation which appear when the medium is composed of diffusing chains rather than a permanent network.

I. Summary of the Current Models

The Doi-Edwards theory deals with the large scale motions of long linear random coil chains in an environment densely filled with the strands of other chains[2, 3)]. The contour of each chain is encompassed by a mesh or lattice of strands, characterized by a mesh size or average lattice spacing which depends only on the polymer concentration and is small compared to overall chain dimensions (Fig. 1). The mesh retards sustained motions transverse to the chain contour but leaves tangential motions essentially unaffected. The chain can explore laterally, but net displacements of the chain as a whole can take place only by diffusion along its own contour. In effect, each chain is confined in a tube. The centerline of the tube tracks the current chain conformation, at least in a coarse-grained sense (Fig. 2). The diameter of the tube, governed by the mesh size, is a measure of the mean span of transverse excursions. The centerline, the primitive path of the chain[2)], is assumed to be a random walk sequence of N steps with step length a (again dependent only on mesh size). The end-to-end vectors of tube and chain coincide, so

$$Na^2 = \langle R^2 \rangle \quad (1)$$

$$Na = L \quad (2)$$

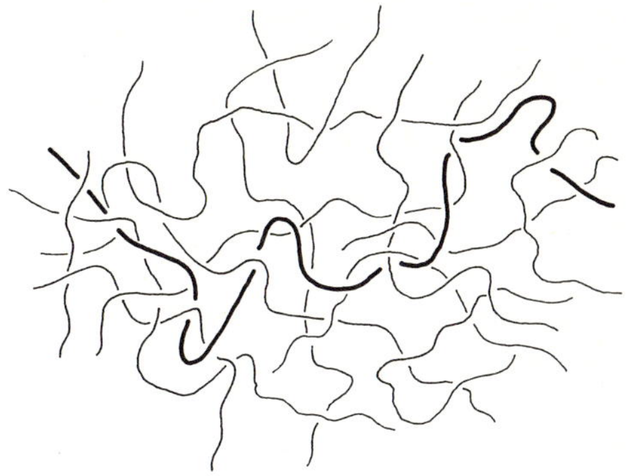

Fig. 1. A polymer chain and its surroundings in an entangled polymer liquid

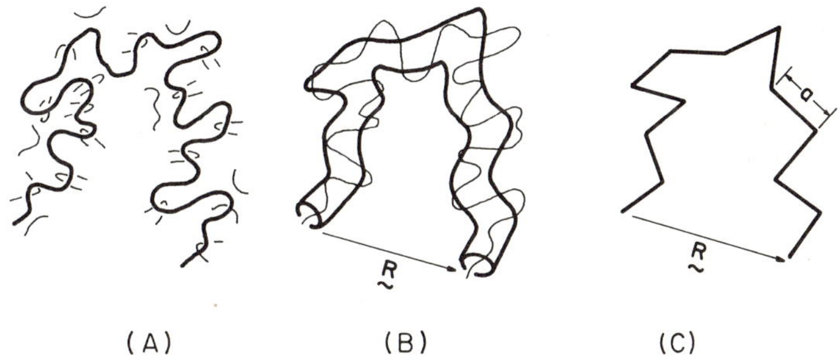

Fig. 2 A–C. Various representations of a polymer chain and its surroundings. The chain and segments of neighboring chains (**A**), the chain in a tube of uncrossable constraints provided by its neighbors (**B**), and the primitive path of a chain among the surrounding constraints provided by neighbors (a = step length of the primitive path; *R* = end-to-end vector of the chain) (**C**)

where L is the path length (not the chain length L_0 which is much larger than L), and $\langle R^2 \rangle$ is the mean-square end-to-end distance of the chains. Both N and L are directly proportional to chain length. Some equilibrium density of chain (monomers/length) is distributed uniformly along the path, and this chain density depends only on the mesh size.

The step length of the primitive path and the tube diameter (the mean span of lateral excursions) are assumed to be independent of chain length and comparable in magnitude. Doi and Edwards take them to be equal, although this is not necessary. One might also expect them to vary with polymer concentration in the same way, but their ratio could conceivably vary with the polymer species. It is important to note, however, that the tube diameter itself is in fact irrelevant except for discussion purposes. Only the primitive path step length a enters the Doi-Edwards derivations for the properties of concern here. It is the quantity which characterizes the topology of the system. Throughout this paper we will reserve the symbol a to mean only the Doi-Edwards step length. The mesh size d characterizes the topology in a new model (Part II).

Center of gravity motion of the chain is governed by its diffusion along the tube. The chain end in the instantaneous direction of motion moves randomly through the mesh as the chain emerges from its current tube. The chain thus creates new primitive path steps with random orientations at the emerging end, and it simultaneously vacates currently occupied steps at the other end. Continuation of this process results in the eventual abandonment of all steps in any initial tube and thus a complete rearrangement of the chain conformation. A step in the primitive path of any initial conformation is abandoned at the first time either of the two chain ends diffuses through it. As long as an initial step survives it preserves its part of the initial conformation and imparts its orientation on whatever portion of the chain that is currently occupying it. The molecule is assumed to behave like a Rouse chain, free to move anywhere but subject to a set of uncrossable spatial constraints which prevent sustained lateral motions. Alternatively, the confining effects of the surrounding chains is represented by a potential centered on the primitive path. In either view, the diffusion coefficient along the tube, D*, and the equilibration time for conformational fluctuations within the tube, τ_e, are given by the Rouse equations:

$$D^* = D_r = \frac{kT}{n_0 \zeta_0} \tag{3}$$

$$\tau_e \approx \tau_r = \frac{\langle R^2 \rangle}{6\pi^2 D_r} \tag{4}$$

where τ is the longest relaxation time of a free Rouse chain, n_0 is the number of monomer units in the molecule and ζ_0 is the friction coefficient per monomer unit[12)].

For times longer than τ_e the effects of reptation by a collection of chains are obtained from solutions of a diffusion equation. Fluctuations in chain density along the tube are omitted; the path length is assumed to be constant, and changes in occupation of path steps take place only by movements of the chain as a whole. Thus, for example, a step located at position x along some initial path ($o < x < L$) will still be occupied at a later time t only if neither chain end has passed through it during that interval. That will be the case only if the chain has never moved farther than x in one direction along the tube or L-x in the other.

Let the center of each chain start at its own origin, and let ξ be the distance from that origin measured along the tube. The fraction of chains located between ξ and $\xi + d\xi$ after time t, $f(\xi, t)d\xi$, is governed by the one dimensional diffusion equation

$$\frac{\partial f}{\partial t} = D^* \frac{\partial^2 f}{\partial \xi^2} \tag{5}$$

with the initial condition

$$f(\xi, o) = \delta(\xi) \text{ (Dirac delta function)} \tag{5a}$$

The absorbing boundary condition selects only those chains whose centers have never moved outside the range $x\text{-}L < \xi < x$:

$$f(x\text{-}L, t) = f(x, t) = o \tag{5b}$$

The solution of Eq. 5 with these boundary conditions is

$$f(\xi, t; x) = \frac{2}{L}\sum_{n=1}^{\infty} \sin\left(\frac{n\pi x}{L}\right) \sin\left(\frac{n\pi}{L}(x - \xi)\right) \exp\left[-\frac{n^2\pi^2 D^* t}{L^2}\right] \tag{6}$$

where $f(\xi, t; x)\, d\xi$ is the fraction of all chains whose centers are located between ξ and $\xi + d\xi$ at time t and for which the center has never been moved outside the range $x\text{-}L < \xi < x$.

From Eq. 6, the fraction of all steps initially located at x which are still occupied after time t is $\int_{x-L}^{x} f(\xi, t; x) d\xi$:

$$F(t; x) = \frac{4}{\pi}\sum_{\substack{n \\ \text{odd}}} \frac{1}{n} \sin \frac{n\pi x}{L} \exp\left[-\frac{n^2\pi^2 D^* t}{L^2}\right] \tag{7}$$

The fraction of all initial steps still occupied at time t is $\frac{1}{L}\int_0^L F(t; x) dx$:

$$F(t) = \frac{8}{\pi^2}\sum_{\substack{n \\ \text{odd}}} \frac{1}{n^2} \exp\left[-n^2 t/\tau_d\right] \tag{8}$$

where τ_d is the disengagement time

$$\tau_d = \frac{L^2}{\pi^2 D^*} \tag{9}$$

From Eq. 1, 2, 4 and 9, $\tau_d \propto M^3$, and

$$\tau_d/\tau_e = 6N \tag{9a}$$

so $\tau_d \gg \tau_e$ for long chains. The progressive disengagement of a chain from an initial path through the surroundings is illustrated in Fig. 3.

A. Non-Mechanical Dynamic Properties

The end-to-end vector correlation function is defined as

$$C(t) = \frac{\langle \mathbf{R}(t) \cdot \mathbf{R}(o) \rangle}{\langle R^2 \rangle} \tag{10}$$

For any chain **R** can be expressed as the sum of its primitive path step vectors $\mathbf{r}_i$:

$$\mathbf{R}(t) = \sum_{i=1}^{N} \mathbf{r}_i(t) = a \sum_{i=1}^{N} \mathbf{u}_i(t) \tag{11}$$

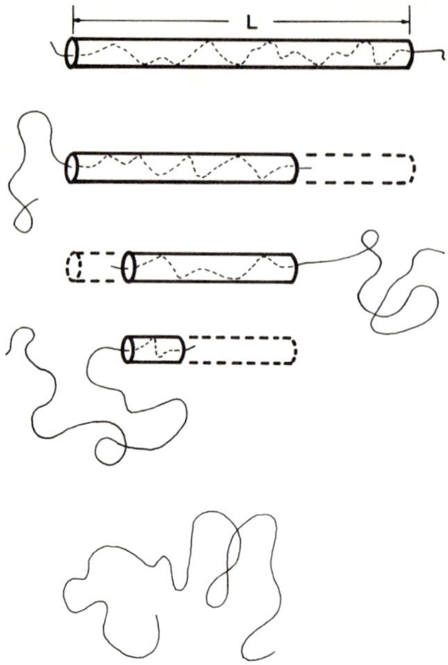

Fig. 3. A reptating chain at various stages in disengagement from its initial tube

where the $\mathbf{u}_i(t)$ are unit vectors parallel to the steps. Since $\langle R^2 \rangle = Na^2$,

$$C(t) = \frac{1}{N}\left\langle \sum_{i=1}^{N}\sum_{j=1}^{N} \mathbf{u}_i(t) \cdot \mathbf{u}_j(o) \right\rangle \tag{12}$$

The primitive path at time t will consist of the initial steps which have survived and the new steps, occupied since t = o. If μ steps have survived there will be μ terms in the double sum for which $\mathbf{u}_i(t) \cdot \mathbf{u}_j(o) = 1$. The remaining terms are products of uncorrelated vectors and will drop out in the ensemble averaging to leave

$$C(t) = \frac{\langle \mu(t) \rangle}{N} \tag{13}$$

Thus, C(t) is just the fraction of all steps which have survived up to time t, so from Eq. 8,

$$C(t) = \frac{8}{\pi^2} \sum_{\substack{n \\ \text{odd}}} \frac{1}{n^2} \exp\left[-n^2 t/\tau_d\right] \tag{14}$$

a result first obtained by de Gennes[1)].

The average movements of reptating chains can be expressed in terms of the diffusion coefficient in the tube. Thus, from the diffusion equation ($t \gg \tau_e$)

$$\langle \xi^2 \rangle = 2D^*t \tag{15}$$

where $\langle \xi^2 \rangle$ is the mean square displacement of chain centers measured along their tubes. Diffusion can also be expressed in terms of an elementary random jump model, displacements along the tube being goverened in this case by the equation for random walks:

$$\langle \xi^2 \rangle = \delta^2 \phi t \tag{16}$$

in which δ is a randomly directed jump distance (+ or – direction along the tube) and ϕ is the jump frequency. Thus,

$$\phi = \frac{2D^*}{\delta^2} \tag{17}$$

The macroscopic diffusion coefficient can be similarly expressed, using the same jump frequency:

$$D = \frac{1}{6} \overline{(\mathbf{\Delta R}_g)^2} \phi \tag{18}$$

but where $\mathbf{\Delta R}_g$ is now the spatial displacement of the center of gravity produced by a jump, and $\overline{(\mathbf{\Delta R}_g)^2}$ is the long time average of $\mathbf{\Delta R}_g \cdot \mathbf{\Delta R}_g$ for reptating chains. Each time a jump occurs the entire chain translates $\pm\,\delta$ along the tube. This is equivalent to transferring a fraction δ/L of the chain from one end to the other. Thus,

$$\mathbf{\Delta R}_g = \pm \frac{\delta}{L} \mathbf{R} \tag{19}$$

and therefore

$$\overline{(\mathbf{\Delta R}_g)^2} = \left(\frac{\delta}{L}\right)^2 \overline{R^2} \tag{20}$$

Since the time average $\overline{R^2}$ for an individual chain becomes equal to the ensemble average $\langle R^2 \rangle$ at long times. Eqs. 17, 18 and 20 combine to give[2)]

$$D = \frac{D^*}{3N} \tag{21}$$

The equivalent expression for the sedimentation coefficient is[22)]

$$s = 3N \frac{kT}{D^*} \tag{22}$$

B. Dynamic Mechanical Properties

Doi and Edwards were the first to exploit the direct connection in reptating systems between the decay in conformational correlation and the relaxation of stress following a step strain[3]. Deformation carries the primitive paths into new conformations, changing the distribution of conformations of the chains and storing free energy in the system. Some decays away rapidly ($t \sim \tau_e$) as the chains equilibrate inside their distorted tubes, but the remainder, associated with chain orientations imposed by the distorted tube conformations, produces a restoring force which decays much more slowly.

It is remarkable that several predictions about linear viscoelastic properties in the terminal region can be obtained without special assumptions about the stress- orientation relationship. Thus, for a simple shear strain γ which is small enough to evoke only a linear response, the shear stress after the rapid equilibratin process can be written

$$c(\tau_e) = G_N^0 \gamma \tag{23}$$

where G_N^0 is the plateau modulus of the system. Assuming then only that the stored energy is distributed uniformly along the chains on average, the subsequent rate of stress relaxation should correspond precisely with the rate of abandonment of primitive path steps. Thus,

$$\frac{\sigma(t)}{\gamma} \equiv G(t) = G_N^0 F(t) \qquad (t > \tau_e) \tag{24}$$

where G(t) is the stress relaxation modulus in the terminal region, and F(t) is the fraction of occupied steps after equilibration which are still occupied at later time t (Eq. 8). From the general equations of linear viscoelasticity[12],

$$\eta_0 = \int_0^\infty G(t)\, dt \tag{25}$$

$$J_e^0 = \frac{1}{\eta_0^2} \int_0^\infty t\, G(t)\, dt \tag{26}$$

With these equations the steady state viscosity and recoverable compliance for sufficiently long chains (so that relaxation processes for $t \lesssim \tau_e$ make negligible contributions to the integrals) can be calculated:

$$\eta_0 = \frac{\pi^2}{12} G_N^0 \tau_d \tag{27}$$

$$J_e^0 = \frac{6}{5 G_N^0} \tag{28}$$

Thus, certain predictions, that $\eta_0 \propto M^3$ and that the product $G_N^0 J_e^0$ is universal, are independent of the form chosen to relate stress and orientation.

Doi and Edwards argue for a particular relationship between stress and orientation[3]. They assume that the primitive path steps deform affinely (Fig. 4) and that the chain segments contained in those steps respond initially like independent Gaussian strands.

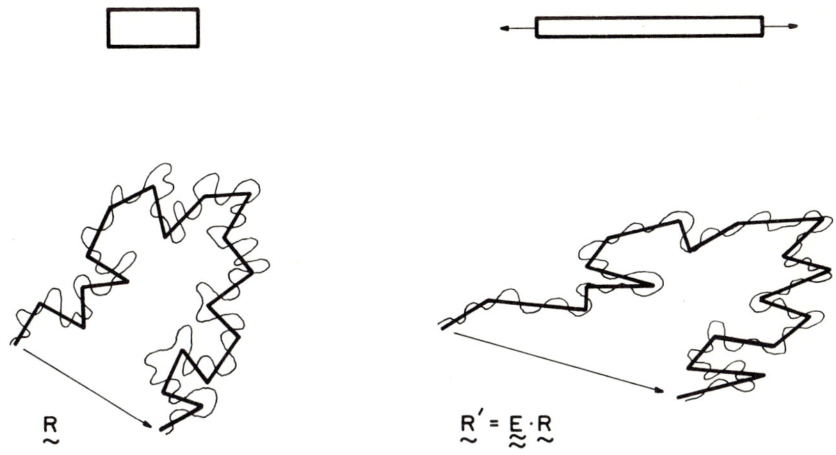

Fig. 4. Distortion of the primitive path by an affine deformation (uniaxial extension)

Thus, very shortly after a step deformation from rest (long enough for local relaxation of the chain) the response is the same as for a Gaussian network with affine junction displacements:

$$\sigma_{\alpha\beta}(o) = 3\nu kT \left\langle \sum_{i=1}^{N} \frac{(\mathbf{E}\cdot\mathbf{r}_i)_\alpha(\mathbf{E}\cdot\mathbf{r}_i)_\beta}{\langle \mathbf{r}_i\cdot\mathbf{r}_i\rangle} \right\rangle = 3\nu kT \left\langle \sum_{i=1}^{N} (\mathbf{E}\cdot\mathbf{u}_i)_\alpha(\mathbf{E}\cdot\mathbf{u}_i)_\beta \right\rangle \tag{29}$$

where $\mathbf{E}$ is the displacement gradient tensor, $\mathbf{u}_i$ is a unit vector parallel to step i in the initial state, and $\langle \mathbf{r}_i \cdot \mathbf{r}_i\rangle = a^2$ is the free mean-square ent-to-end distance of the chain segment contained in step i.

During equilibration ($t < \tau_e$) the chain density along the path, changed locally by the stretching and compressing of steps according to their initial orientations, returns to its equilibrium value (Fig. 5). This affects the stress in two ways. First, the path length after deformation is $a \sum_{i=1}^{N} |\mathbf{E}\cdot\mathbf{u}_i|$, which is equal to $L\langle|\mathbf{E}\cdot\mathbf{u}|\rangle$ for each chain if N is large. Equilibration of chain density along the path returns the length to L, the equilibrium value. Since $\langle|\mathbf{E}\cdot\mathbf{u}|\rangle$ is always greater than unity for a step deformation from rest[3)], the chain ends retract during equilibration, abandoning a fraction $1 - \langle|\mathbf{E}\cdot\mathbf{u}|\rangle^{-1}$ of the initial steps. Second, the amount of chain in each remaining step becomes proportional to its deformed length, $a|\mathbf{E}\cdot\mathbf{u}_i|$. Thus, $\langle \mathbf{r}_i\cdot\mathbf{r}_i\rangle$ changes from a^2 to $a^2|\mathbf{E}\cdot\mathbf{u}_i|$. The combination of these two effects gives, after equilibration,

$$\sigma_{\alpha\beta}(\tau) = \frac{3\nu kT}{\langle|\mathbf{E}\cdot\mathbf{u}|\rangle} \left\langle \sum_{i=1}^{N} \frac{(\mathbf{E}\cdot\mathbf{u}_i)_\alpha(\mathbf{E}\cdot\mathbf{u}_i)_\beta}{|\mathbf{E}\cdot\mathbf{u}_i|} \right\rangle \tag{30}$$

If N is large, the quantity $\sum_{i=1}^{N}(\ )$ in Eq. 30 can be replaced by $N\langle(\ )\rangle$. Thus, the stress components at three stages in time – shortly after deformation, just after equilibration,

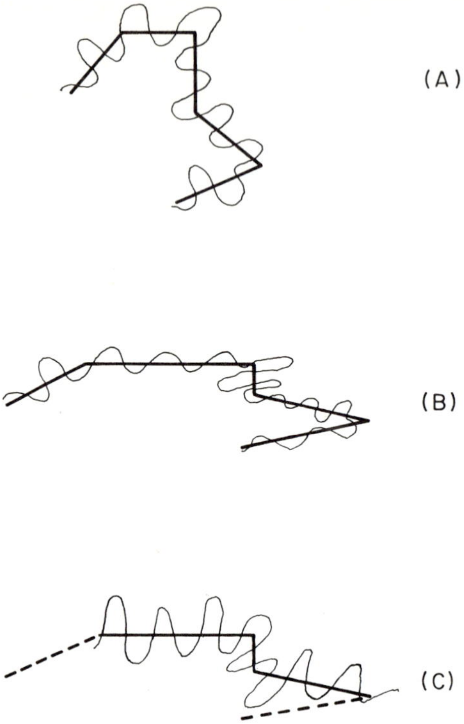

Fig. 5 A–C. Changes in local density of chain along the path during equilibration. A chain and its primitive path before deformation (**A**), the same chain and path after deformation but before equilibrium (**B**), and the equilibrated chain after deformation, showing the disengagement from part of the path when the density of chain along the path returns to equilibrium (**C**)

and during subsequent reptation out of the distorted tubes – are given by the following expressions[3)]:

$$\sigma_{\alpha\beta}(t) = 3\nu NkT \langle (\mathbf{E}\cdot\mathbf{u})_\alpha (\mathbf{E}\cdot\mathbf{u})_\beta \rangle \qquad \tau_{tr} < t < \tau_e \tag{31}$$

$$\sigma_{\alpha\beta}(t) = \frac{3\nu NkT}{\langle |\mathbf{E}\cdot\mathbf{u}| \rangle} \left\langle \frac{(\mathbf{E}\cdot\mathbf{u})_\alpha (\mathbf{E}\cdot\mathbf{u})_\beta}{|\mathbf{E}\cdot\mathbf{u}|} \right\rangle \qquad t \sim \tau_e \tag{32}$$

$$\sigma_{\alpha\beta}(t) = \sigma_{\alpha\beta}(\tau_e)F(t) \qquad t > \tau_e \tag{33}$$

where τ_{tr} is the chain length independent time, characterizing the transition region[12)].

The brackets $\langle\ \rangle$ indicate an average over all directions of the unit vector **u**. In polar coordinates $\mathbf{u} = [\cos\theta, \sin\theta\cos\phi, \sin\theta\sin\phi]$ and

$$\left\langle (\ ) \right\rangle = \frac{1}{4\pi} \int_0^{2\pi} \int_0^{\pi} (\ ) \sin\theta d\theta d\phi \tag{34}$$

For a simple shear strain

$$\mathbf{E} = \begin{bmatrix} 1 & \gamma & 0 \\ 0 & 1 & 0 \\ 0 & 0 & 0 \end{bmatrix} \tag{35}$$

If γ is small, Eqs. 31 and 32 give

$$\frac{\sigma_{21}(t)}{\gamma} = G(t) = \nu NkT \qquad t < \tau_e \qquad (36)$$

$$\frac{\sigma_{21}(t_e)}{\gamma} = G_N^0 = \frac{4}{5}\nu NkT \qquad t \sim \tau_e \qquad (37)$$

The last expression, which follows directly from the Doi-Edwards expression for stress, provides a link between mechanical behavior and the center-of-gravity diffusion coefficient. The tube parameter N can be eliminated between Eqs. 21 and 37, and D* can be estimated from independent experiments as discussed earlier. Confirmation of this relationship[11)], the observed separability of strain and time dependences in step strain relaxation experiments[23)] as required by Eq. 33:

$$\frac{\sigma_{21}(t, \gamma)}{\gamma} = G(t, \gamma) = h(\gamma)\, G(t) \qquad (t > \tau_e) \qquad (38)$$

and agreement of the experimental $h(\gamma)$ with the predictions of Eq. 32 in many systems[24)] provide strong support for the reptation picture generally as well as for the explicit expression for the stress proposed by Doi and Edwards. Departures of experimental data from Eqs. 27 and 28 for η_0 and J_e^0 (Fig. 6), however, imply the existence of relaxation processes which compete with reptation in various situations, as discussed in the introduction.

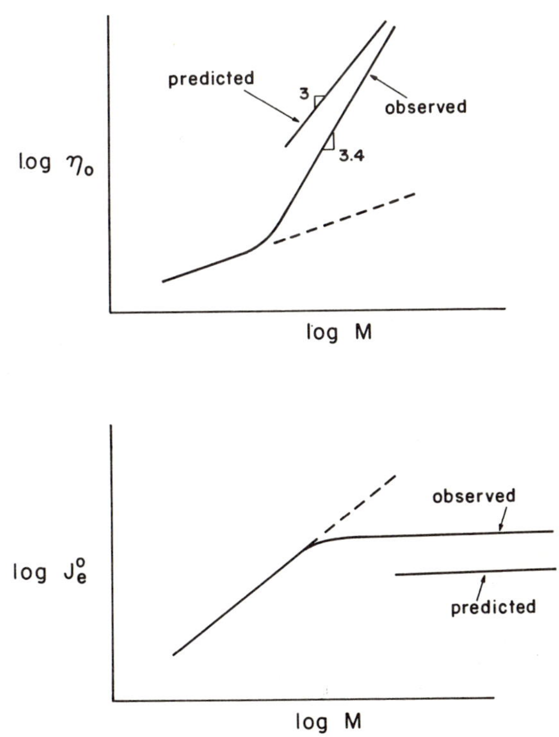

Fig. 6. Comparisons of viscosity and recoverable compliance predictions by the Doi-Edwards theory with experimental observations. The predicted η_0 is too large, but its chain length dependence is slightly weaker than observed. The predicted J_e^0 is too small, but independent of chain length as observed. The dashed lines indicate predictions of the Rouse model

C. Equilibrium Elasticity in Entangled Networks

In addition to these results for entangled chain liquids, the Doi-Edwards formula for stress implies a specific form for entanglement contributions to the equilibrium stored energy function in crosslinked networks, which can be developed as follows. Immediately following an affine deformation the stress from entanglements is given by Eq. 29. For a permanent network, the primitive path length $L\langle|\mathbf{E}\cdot\mathbf{u}|\rangle$ and the number of primitive path steps of each network strand must remain unchanged during equilibration. The density of chain along the path eventually becomes uniform however by redistribution among steps of the same network strand according to the changes in their individual lengths (Fig. 7).

Thus, $\langle\mathbf{r}_i\cdot\mathbf{r}_i\rangle$ changes to $\langle\bar{\mathbf{r}}\cdot\mathbf{r}_i\rangle|\mathbf{E}\cdot\mathbf{r}_i|/\langle|\mathbf{E}\cdot\mathbf{r}_i|\rangle$, and the equilibrium stress for large N is given by

$$\sigma_{\alpha\beta} = 3\,\nu NkT\,\langle|\mathbf{E}\cdot\mathbf{u}|\rangle\left\langle\frac{(\mathbf{E}\cdot\mathbf{u})_\alpha(\mathbf{E}\cdot\mathbf{u})_\beta}{|\mathbf{E}\cdot\mathbf{u}|}\right\rangle \tag{39}$$

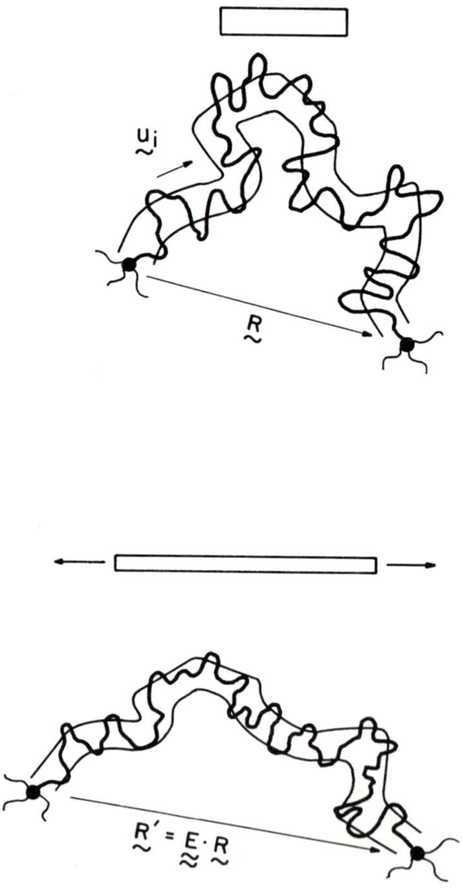

Fig. 7. The Doi-Edwards model applied to the strands of a network. The end-to-end vector of each strand changes according to the macroscopic deformation, in this case a uniaxial extension. The path length increases, and the chain density per unit path length decreases with deformation

It is a straightforward exercise to show that Eq. 39 is the expression for stress components in an elastic incompressible solid with a stored energy function W (energy/volume) given by

$$W = \frac{3\nu NkT}{2} [\langle |\mathbf{E} \cdot \mathbf{u}| \rangle^2 - 1] \tag{40}$$

If entanglements acted like ordinary crosslinks ($\nu N/2$) per unit volume) the stored energy function would be given by the usual expression for tetrafunctional phantom networks with the spatial fluctuations of junctions suppressed[25)]

$$W = \frac{3\nu NkT}{2} [\langle |\mathbf{E} \cdot \mathbf{u}|^2 \rangle - 1] = \nu NkT \frac{\alpha_1^2 + \alpha_2^2 + \alpha_3^2 - 3}{2} \tag{41}$$

where the α's are principal stretch ratios.

Equations 40 and 41 involve different averages of the elemental stretches and give slightly different shear moduli for small deformations, 4/5 νNkT and νNkT respectively. However, the shapes of the stress-strain curves would be difficult to distinguish experimentally (see Sect. II C.). Thus, although this rather natural application of the Doi-Edwards model to permanent networks leads directly to a simple additive contribution of trapped entanglements to the initial modulus, as observed experimentally ($G_e^{max} \approx G_N^0$ [26)]), it does not account for the quite substantial departures from the neo-Hookean form (Eq. 41) which are commonly observed at even moderate deformations[27, 28)].

II. Chains in Topologically Invariant Surroundings

We propose here a model which incorporates both reptative motions and fluctuations in length of the occupied path. Part III offers a simple extension to include also the effect of finite lifetimes of constraints in polymer liquids. Other workers[29, 30)] have proposed alternative ways to handle the reptation idea. These will be considered briefly in the discussion (Part IV).

A. General Features of the Model

Consider a random-walk chain of n_0 monomer units in a medium which is densely filled with the contours of other chains. For the moment take the ends of the test chain to be fixed. Let its surroundings be represented by a permanently connected rigid lattice of uncrossable lines enveloping the chain contour. We assume that the effect of this obstacle lattice on the conformations of the chain is specified simply by a distance scale, the mesh size d, as follows. Pieces of the chain which have a mean-square end-to-end distance $\langle r^2 \rangle$ much smaller than d^2 can explore all conformations with the same probability as free chains of the same length. For longer pieces, the presence of the obstacles (and the fact that the pieces are connected in a definite sequence between the fixed end points of the chain) begins to reduce conformational accessibility relative to free chains of the same size. Pieces for which the free chain $\langle r^2 \rangle$ is much larger than d^2 have many fewer conformations, or, more precisely, the probability of most conformations is greatly reduced for the long pieces relative to the free chain. Where $\langle r^2 \rangle \sim d^2$ the available conformations are assumed to correspond roughly to those of an "almost free" chain, meaning one in which only the average orientation of the end-to-end vector is strongly influenced by the obstacles.

The mesh size d is assumed to depend only on the effective spacing of obstacles in the medium and thus on the polymer concentration alone. We do not mean to imply that d would necessarily be of the order of the distance between chains (a few Angstroms in undiluted systems). Its effective value should be somewhat larger than this, since d must also reflect a certain amount of local freedom for mutual rearrangement of neighbors in real systems. Nevertheless, like the primitive path step length a of Doi and Edwards[2)], d should be independent of the large scale molecular structure.

The same considerations should also apply for long chains without fixed ends. In this case all conformations are accessible eventually, but, for times which are small compared to those required for a complete rearrangement of conformation, the definition of mesh size and comments about its influence on the accessible local conformations should still be valid.

To model the effects of large scale structure on the properties we will represent each chain by a random-walk sequence of m_0 pieces or *primitive segments* in which $\langle r^2 \rangle = d^2$ for each segment. Thus m_0 d is the sequence length, and, for a collection of chains of the same length,

$$\langle R^2 \rangle = m_0 d^2 \tag{42}$$

where $\langle R^2 \rangle$ is the mean-square end-to-end distance. We assume m_0 is large, but still small compared to n_0. If the chain ends are fixed, the sequence of segments is free to assume any conformation on condition that it can be reached without crossing any of the obstacle lines. This means that primitive segments must always occupy all positions along one particular locus among the obstacles, namely the Doi-Edwards primitive path[2)].

The primitive path is established uniquely for each chain when its ends are fixed (or when crosslinks are added in the case of a network). The number of primitive segments required to occupy the primitive path is also fixed. Suppose this number is m ($o < m < m_0$) for a given chain with fixed ends (Fig. 8.) Then the remaining $m_0 - m$ segments represent surplus length. At any moment the surplus is arranged in some manner along the primitive path in the form of local loops, projecting from the path at up to $m + 1$ possible locations and not enclosing any obstacles. Any primitive segment i ($i = 1, ---, m_0$) spends part of its time as a member of a sequence occupying the primitive path (the path population) and the remainder as a member of a loop sequence (the surplus population) infiltrating the surroundings. Local motions allow free exploration of locations and orientations by segment i, subject only to the conditions that its end-to-end vector $\mathbf{r}_i$ joins the ends of adjacent segments $i - 1$ and $i + 1$ and that all sequence conformations share the same primitive path. The Doi-Edwards model suppresses the effects of these explorations; the present model includes them, although in a rather simplified fashion.

The primitive path length L_m of a chain which has m path segments is

$$L_m = md \tag{43}$$

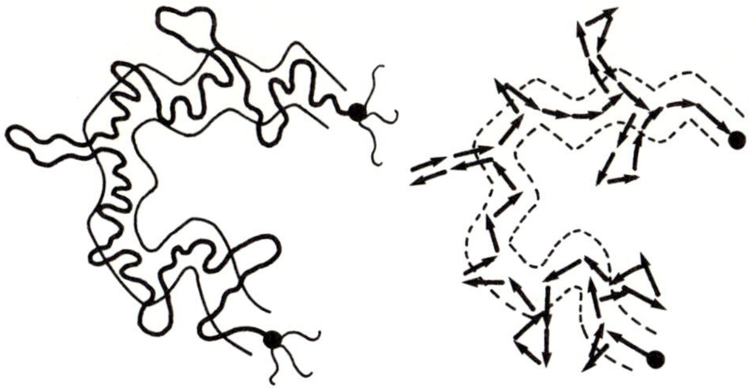

Fig. 8. A network strand showing path-occupying and surplus portions as represented by the primitive segment model. The length of the arrows indicated the mesh size in the system

The fraction of such chains is governed by some probaility density P_m (see below). We will assume for most purposes that m is closely distributed about its average

$$\langle m \rangle = \sum_{m=0}^{m_0} m P_m = N \tag{44}$$

where N is the number of primitive path steps in the Doi-Edwards model. If we assume that primitive paths can be regarded as random walks with some equivalent step length b, then (with $L \equiv Nd$),

$$\langle R^2 \rangle = b N d \tag{45}$$

and, from Eqs. 42 and 44.

$$b = \frac{m_0}{N} d \tag{46}$$

Thus, the effective primitive path step length and the mesh size differ in this model according to the average amount of surplus length along the path. The mesh size d is the fundamental parameter defining the topology; the distinction is useful however only when the effect of fluctuations in the amount of surplus length is important.

If the ends of each chain are fixed to the medium the length of each primitive path changes with deformation. When m is large, the path length for a constant volume deformation $\mathbf{E}$ increases from md to md $\langle |\mathbf{E} \cdot \mathbf{u}| \rangle$, giving m $\langle |\mathbf{E} \cdot \mathbf{u}| \rangle$ primitive segments in the path population and $m_0 - m \langle |\mathbf{E} \cdot \mathbf{u}| \rangle$ surplus segments. An individual segment (except one near the chain ends) has a range of path locations to sample. When in the path it is one of a primitive segment sequence running along the path, and we assume its end-to-end vector is parallel to the local path tangent at that location s along the path

$$\mathbf{r}_i \cdot \mathbf{u}(s) = d \tag{47}$$

When it is in the surplus population (a member of a loop sequence) we assume that its end-to-end vector is uncorrelated with path direction and independent of the deformation.

Thus, the distribution of orientations for an individual segment depends on the fraction of time it spends in the path and on the distribution of path orientations along the span it can easily sample. If the span is large enough the primitive segment will sample the full distribution of orientations of primitive path tangent vectors. In the undeformed state this distribution is random; in the deformed state it is the distribution corresponding to affine displacement of a set of random tangent vectors.

B. Path Length Probability Function P_m

Let P_m be the probability that a free chain with m_0 primitive segments occupies m primitive path steps. This is the same as the probability that a chain has a total of $S = m_0 - m$ surplus segments, i.e., that $m_0 - m$ segments are parts of excusions or loops which do not enclose any obstacles. De Gennes used a regular obstacle lattice to show that q(S), the probability of forming a closed random walk sequence of S segments without obstacle enclosure, is approximately exponential[17]:

$$q(S) = \alpha e^{-\alpha S} \tag{48}$$

Evans, in a numerical simulation exploring primitive paths for random walks which interpenetrate an obstacle lattice, also found an exponential distribution of surplus sequence lengths[31]. A term similar to Eq. 48 appears in the Doi-Edwards expression for the distribution of chain along a tube defined by a quadratic confining potential (Eq. A. 6, Ref. 3).

Equation 48 suggests that surplus primitive segments might be equivalently regarded as having excess free energy relative to path segments, and that this free energy is practically independent of the details of sequence arrangement. Thus, the net conversion of one surplus segment to a path segment in a chain without fixed ends increases the path length by one segment. The configurational entropy of the chain is changed by the difference between the entropy of a primitive segment choosing any direction as it emerges from one end of the original path and the entropy of a primitive segment which is part of a closed loop sequence. In the latter case its orientation is conditional to some extent on the orientations of other segments. Accordingly, one might expect that, for a chain with m_0 primitive segments, the path length probability density has the general form

$$P_m = \frac{K_m(m_0)\, e^{-\alpha(m_0 - m)}}{\sum_{m=0}^{m_0} K_m(m_0)\, e^{-\alpha(m_0 - m)}} \tag{49}$$

where $K_m(m_0)$ is the number of distinguishable arrangements of $m_0 - m$ surplus segments among $m + 1$ possible locations.

A simple expression for P_m is obtained by taking p to be the probability that a primitive segment selected at random is at that moment a member of the surplus population and by assuming that the simultaneous probabilities for segments on the same chain are uncorrelated. This stipulation of "maximum randomness" produces directly the expected exponential form for q(S) since the probability of finding a sequence of S surplus segments ($S \ll m_0 - m$) is proportional, to p^S. It also leads to a binomial distribution for P_m because the combinatorial term $K_m(m_0)$ becomes simply $m_0!/m!(m_0 - m)!$:

$$P_m = \frac{m_0!}{m!(m_0 - m)!} p^{m_0 - m}(1 - p)^m \tag{50}$$

where α in Eq. 48 is $\ln \frac{1-p}{p}$.

The mean and variance for this distribution are

$$\langle m \rangle = m_0(1 - p) = N \tag{51 a}$$

$$\langle m^2 \rangle - \langle m \rangle^2 = m_0 p(1 - p) = \langle m \rangle p \tag{51 b}$$

When $m_0 \gg \langle m \rangle \gg 1$, the distribution near the mean is approximately Gaussian:

$$P(m) = \left(\frac{1}{2\pi\langle m \rangle p}\right)^{1/2} \exp\left[-\frac{(m - \langle m \rangle)^2}{2\langle m \rangle p}\right] \tag{52}$$

and, for small m and large m_0,

$$P_m = \left(\frac{\langle m \rangle}{p}\right)^m \frac{1}{m!} \exp\left[-\frac{\ln(1/p)}{1 - p}\langle m \rangle\right] \tag{53}$$

This assumption of maximum randomness is admittedly crude, but in the absence of more detailed information we will use it when an explicit form for P_m is required. Other assumptions are possible. For example, in the primitive segment model the surplus segments in each of the m + 1 possible local loops must occur in pairs in order to be self-cancelling. For that case the combinatorial term $K_m(m_0)$ is $(m_0 + m)/2!/m!(m_0 - m/2)!$ for m even and o for m odd (Appendix I). This leads to a different expression for P_m, but the approximation equations for large m_0 (Eqs. 52 and 53) have the same forms.

C. Equation for the Stress and Rubber Elasticity

Deformation in systems where the chain ends are permanently anchored in the medium alters the distribution of orientations of segments occupying the primitive path, and it reduces the number of surplus segments in each chain. There is a free energy change associated with each of these effects. For $\langle m \rangle$ large the results do not depend on strand length distribution, so for simplicity we consider all chains to have m_0 segments.

Consider the first contribution from segments which occupy the primitive path. Earlier we proposed the approximation that individual primitive segments are not stretched

($|\mathbf{r}_i| = d$) even for moderate deformations. Instead, segments are withdrawn from the surplus population as the path length increases, maintaining a constant number of path-occupying segments per unit length of path. A chain with m path segments before deformation has m $\langle|\mathbf{E}\cdot\mathbf{u}|\rangle$ path segments after deformation. A position along the path with unit tangent vector $\mathbf{u}$ in the undeformed state has orientation $\mathbf{u}' = \mathbf{E}\cdot\mathbf{u}/|\mathbf{E}\cdot\mathbf{u}|$ after deformation. The end-to-end vector of a segment occupying that position in the deformed state has the same direction (Eq. 47). The stress components according to Eq. 29 (with $\langle r_i\cdot r_i\rangle = d^2$, $(\mathbf{E}\,\mathbf{r}_i)_\alpha = \frac{(\mathbf{E}\cdot\mathbf{u}_i)_\alpha}{|\mathbf{E}\,\mathbf{u}_i|}$, etc.) are thus given by

$$\sigma_{\alpha\beta} = 3\nu kT \sum_{m=0}^{m_0} P_m \left\langle \sum_{i=1}^{m\langle|\mathbf{E}\cdot\mathbf{u}|\rangle} \frac{(\mathbf{E}\cdot\mathbf{u}_i)_\alpha(\mathbf{E}\cdot\mathbf{u}_i)_\beta}{|\mathbf{E}\cdot\mathbf{u}_i|^2} \right\rangle \tag{54}$$

in which i labels those segments occupying the primitive path of a chain at that moment and ν is the number of chains (network strands) per unit volume. Equation 54 is identical to the expression for stress in a system of rigid rods with the same distribution of orientations (see for example Eq. 11.3-2 without flow terms in Ref. 31 a). This is of course not surprising, since the model assumes orientation without stretching of the primitive segments.

For large m_0 the summation over i in Eq. 54 can be written as an average over all orientations:

$$\sigma_{\alpha\beta} = 3\nu kT\langle|\mathbf{E}\cdot\mathbf{u}|\rangle \left(\sum_{m=0}^{m_0} m\,P_m\right) \int \frac{(\mathbf{E}\cdot\mathbf{u})_\alpha(\mathbf{E}\cdot\mathbf{u})_\beta}{|\mathbf{E}\cdot\mathbf{u}|^2}\,d\mathbf{v} \tag{55}$$

where $d\mathbf{v}$ is the fraction of segments with orientation vector $\mathbf{E}\cdot\mathbf{u}/|\mathbf{E}\cdot\mathbf{u}|$ in the deformed state. The path of each chain is stretched in some places and shortened in others by the deformation. Since the number of path segments per unit path length is independent of deformation those portions of the undeformed path which have lengthened will acquire more segments, and those which have shortened will lose segments. The fraction of segments occupying portions of path which had orientation $\mathbf{u}$ in the undeformed state will thus be

$$d\mathbf{v} = \frac{|\mathbf{E}\cdot\mathbf{u}|}{\langle|\mathbf{E}\cdot\mathbf{u}|\rangle}\,d\mathbf{u} \tag{56}$$

so the orientational contribution to stress is

$$\sigma_{\alpha\beta} = 3\nu NkT \left\langle \frac{(\mathbf{E}\cdot\mathbf{u})_\alpha(\mathbf{E}\cdot\mathbf{u})_\beta}{|\mathbf{E}\cdot\mathbf{u}|} \right\rangle \tag{57}$$

in which Eq. 44 has been used, and $\langle|\mathbf{E}\cdot\mathbf{u}|\rangle$ has cancelled from numerator and denominator. The last term in Eq. 57 is the stress for the stored energy function $\langle|\mathbf{E}\cdot\mathbf{u}|\rangle$ (see Eqs. 39, 40), so

$$W_{-1} = 3\nu NkT[\langle|\mathbf{E}\cdot\mathbf{u}|\rangle - 1] \tag{58}$$

and W_1 is the orientational contribution to the free energy of deformation.

The second contribution comes from the change in apportionment of path and surplus segments with deformation. A chain with m path segments at rest has m $\langle|\mathbf{E}\cdot\mathbf{u}|\rangle$ path segments after deformation. If the network was formed from an equilibrium population of chains, the number of distinguishable arrangements of m_0 segments to give m primitive path steps, Ω_m, is proportional to P_m. Deformation changes m to $\langle|\mathbf{E}\cdot\mathbf{u}|\rangle$ m. From the Boltzmann equation the change in configurational entropy per unit volume is

$$\Delta S = k\nu \sum_{m=0}^{m_0} P'_m \quad \ln \frac{P_m}{P'_m} \tag{59}$$

when P'_m is the fraction of chains with m path segments in the deformed state and P_m is the fraction which have m path segments at equilibrium (for a parallel example see p. 467-8 in Ref. 31 b). Since the network was formed at equilibrium, $P'_m = P_{m/\varkappa}$, where $\varkappa = \langle|\mathbf{E}\cdot\mathbf{u}|\rangle$. Thus, with Eq. 52 for P_m

$$\frac{\Delta S}{k\nu} = -\int_{-\infty}^{\infty} P\left(\frac{m}{\varkappa}\right)\left[\frac{(m - \langle m\rangle)^2 - \left(\frac{m}{\varkappa} - \langle m\rangle\right)^2}{2\langle m\rangle p}\right]\frac{dm}{\varkappa} \tag{60}$$

which integrates to give (with $W = F = -TS$ and $N = \langle m\rangle$) the second contribution:

$$W_2 = \frac{\nu NkT}{2p}\left[\langle|\mathbf{E}\cdot\mathbf{u}|\rangle - 1\right]^2 + \nu kT\left[\langle|\mathbf{E}\cdot\mathbf{u}|\rangle^2 - 1\right] \tag{61}$$

The total free energy of deformation per unit volume is the sum of W_1 and W_2. For large N, the entanglement contribution to network elasticity becomes

$$W = 3\nu NkT\left[\langle|\mathbf{E}\cdot\mathbf{u}|\rangle - 1\right] + \frac{\nu NkT}{2p}\left[\langle|\mathbf{E}\cdot\mathbf{u}|\rangle - 1\right]^2 \tag{62}$$

This result differs somewhat from the expression obtained using the Doi-Edwards model (Eq. 40), and it gives a larger departure from neo-Hookean behavior for uniaxial extension (Appendix II and Fig. 9). In the limit of small deformations the entire contribution to stress comes from the first term in Eq. 62. The entanglement contribution to the infinitesimal shear modulus is precisely the same as the Doi-Edwards expression for the plateau modulus (Eq. 37)

$$G_0 = \frac{4}{5}\nu NkT = G_N^0 \qquad \text{(network)} \tag{63}$$

although now $N = \langle R^2\rangle(1-p)/d^2$ (Eqs. 42 and 51 a) instead of $\langle R^2\rangle/a^2$ (Eq. 1) in the Doi-Edwards formulation.

The first term in Eq. 62 is independent of the particular form used to represent P_m. The strain dependence of the second term ($\propto [\langle|\mathbf{E}\cdot\mathbf{u}|\rangle - 1]^2$) depends only on the Gaussian form used to approximate the distribution of m about the mean. The Gaussian

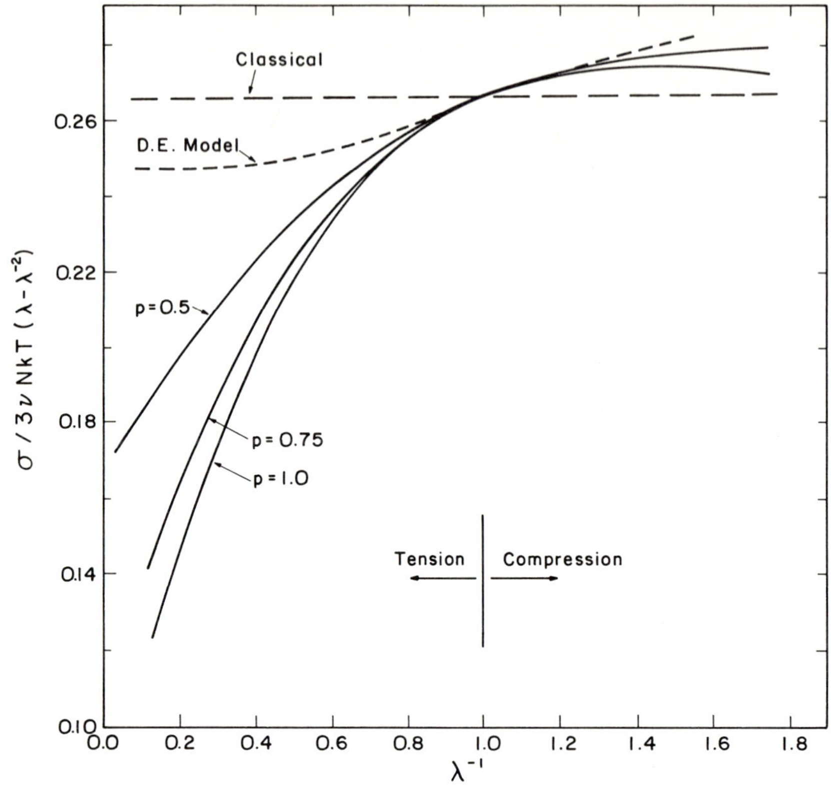

Fig. 9. Equilibrium stress-strain behavior of entangled networks in uniaxial extension and compression. The solid lines (p = 0.5, 0.75, 1.00) were calculated for the primitive segment model (Eq. 62 and II-5). The *short-dash* line is the Doi-Edwards model (Eq. 40 and II-11). The *long-dash* line is the affine Gaussian network model (Eq. 41 and II-12) adjusted to have the same initial modulus

form is expected for large m_0, regardless of the particular choice of P_m. Only the multiplier 1/2 p depends directly on the use of the maximum randomness assumption (Eq. 50); for the paired distribution (Appendix II, that multiplier is $(1 + p)^2/8\,p$. For arbitrary P_m it is $\langle m\rangle/2(\langle m^2\rangle - \langle m\rangle^2)$. Equation 62 must of course break down for strains which are large enough to deplete the surplus segment population appreciably since further deformation would require stretching of the primitive segment lengths.

If the chain ends were released, distribution of surplus segments and the occupied path length would equilibrate (in time τ_e (Eq. 4)). The stress contribution from the second term in Eq. 62 for networks would therefore be lost. The stress from the first term would be reduced by the factor $\langle|\mathbf{E}\cdot\mathbf{u}|\rangle$ because parts of the initial paths would be abandoned as the chains return to an equilibrium set of path lengths[3]. The equation for stress at τ_e then becomes (from Eq. 57)

$$\sigma_{\alpha\beta}(\tau_e) = \frac{3\,\nu NkT}{\langle|\mathbf{E}\cdot\mathbf{u}|\rangle}\left\langle\frac{(\mathbf{E}\cdot\mathbf{u})_\alpha(\mathbf{E}\cdot\mathbf{u})_\beta}{|\mathbf{E}\cdot\mathbf{u}|}\right\rangle \tag{64}$$

which is identical to the Doi-Edwards expression (Eq. 32) for unattached chains. Equation 32 or 64 should apply to both linear and star molecules. Equilibration of path length occurs rapidly everywhere since all parts of the chain communicate directly with a free end. It should not apply for contributions from the interior elements of molecules which contain more than one long-chain branch point.

D. Disengagement by Fluctuations in Path Length

Although the total number of primitive segments in a chain is constant, the partitioning between path and surplus populations can change with time if the chain has a free end. The primitive path lengthens and shortens as the path population increases and decreases: the path length "breathes" as the populations fluctuate (Fig. 10). This provides a mechanism for disengagement (tube leakage) which may compete with simple

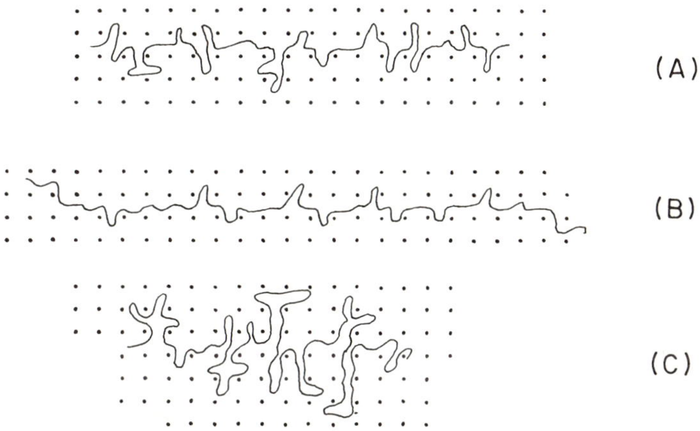

Fig. 10 A–C. Variations in path length from fluctuations in the path and surplus segment populations of an unattached linear chain. An "average" apportionment between path-occupying and surplus primitive segments (**A**), an abnormally small surplus segment population (**B**), and an abnormally large surplus segment population (**C**). The *dots* are the uncrossable obstacles in this two-dimensional representation

reptation or even become the dominant relaxation mechanism if reptation is somehow suppressed.

1. Tethered Chains

Consider a long chain fixed at one end in a medium of obstacles. Being tethered, the chain cannot undergo a pure reptation, but the length of its primitive path can vary with time. Thus, as time passes the number of primitive segments in the path will change as Brownian motion redistributes the path and surplus populations. Sometimes the path length will increase due to a momemtary reduction in surplus segments,and the free end will move out along a random path among the obstacles. Sometimes it will decrease due

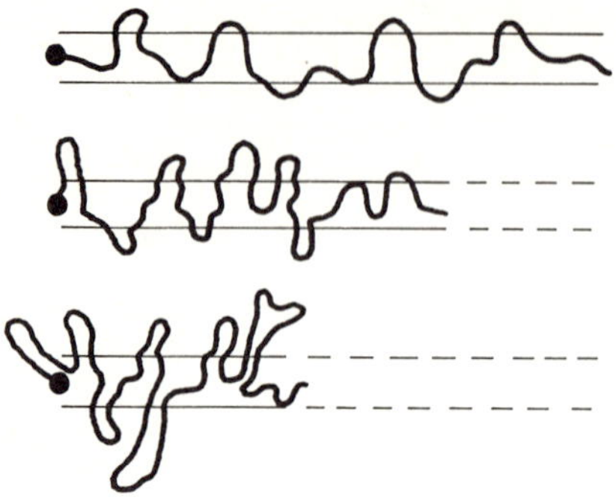

Fig. 11. Abandonment of an initial conformation for a tethered chain by successively larger fluctuations in the surplus segment population

to an increase in surplus segments, and the free end will pull back, disengaging the chain from part of its previously occupied path. Such repeated breathing motions of the primitive path will result in the progressive abandonment of any initially occupied path through the surrounding obstacles. In effect, the initial path of a tethered chain evaporates from the free end inwards even though pure reptation is suppressed (Fig. 11).

Let m(t) be the time-dependent number of path segments for one of the chains, with m(o) as the initial value. Label the initial path positions i = 1,2,---, m(o), beginning at the fixed end. Disengagement from the initial path step at position i then occurs the first time m(t) is equal to i − 1, i.e., the first time the free end of that chain passes through position i along its initial path. The values of m(t) and m(t′) will be highly correlated if the time interval t′ − t = Δt is very small. As Δt increases the correlation should decay roughly as

$$\frac{\langle (m(t+\Delta t)-\overline{m})\,(m(t)-\overline{m})\rangle}{\langle (m(t)-\overline{m})^2\rangle} \propto e^{-\Delta t/\tau_e} \qquad (65)$$

where $\overline{m} = \langle m \rangle = N$, the mean number of path segments (Eq. 44), and τ_e is the equilibration time for the conformational fluctuations of the chain. We assume τ_e in this case corresponds to the longest Rouse relaxation time for a tethered chain, which is easily shown to be (Appendix III)

$$\tau_r = 4\,(\tau_r)_0 \qquad (66)$$

where $(\tau_r)_0$ is the relaxation time for a free chain of the same length (Eq. 4).

Thus, for sampling intervals longer than τ_e, the values of m(t) and m(t + Δt) will be virtually uncorrelated, and the successive path length probabilities can be estimated directly from the equilibrium distribution P_m. We assume that the successive abandonment of initial paths, and thus the stress relaxation modulus, can be adequately described for a collection of tethered chains by the use of P_m and a minimum sampling interval τ_e, as follows.

The probaility that a tethered chain selected at random occupies at least n path steps is

$$Q_n = \sum_{m=n}^{m_0} P_m \qquad (67)$$

Suppose the chain is sampled j times with a sampling interval τ_e. The probability it occupies at least n steps at all these samplings is Q_n^j. Thus, for ν chains per unit volume, the number of steps labelled n (counting from the fixed end) which are initially occupied, and which have not been abandoned after j subsequent samplings, is $\nu\, Q_n^{1+j}$. The total number of steps per unit volume which have survived is therefore $\nu \sum_{n=1}^{m_0} Q_n^{1+j}$. With sampling interval τ_e, the index j is just t/τ_e. The fraction of all steps which survive beyond time t is therefore

$$F(t) = \frac{1}{N}\sum_{n=1}^{m_0} Q_n^{1+t/\tau_e} \qquad \text{(tethered chains)} \tag{68}$$

The contribution of each surviving step to the shear stress relaxation modulus is 4/5 kT (see Eq. 63), so for tethered chains,

$$G(t) = \frac{4}{5}\,\nu kT \sum_{n=1}^{m_0} Q_n^{1+t/\tau_e} = \frac{4}{5}\,\nu kT \sum_{n=1}^{m_0} Q_n \exp\left[-\left(\ln\frac{1}{Q_n}\right)\frac{t}{\tau_e}\right] \tag{69}$$

The expression for plateau modulus ($G_N^0 = G(o)$ in Eq. 69) is the same as that for free linear chains:

$$G_N^0 = \frac{4}{5}\,\nu NkT \tag{70}$$

since $\sum_{n=1}^{m_0} Q_n = \sum_{m=0}^{m_0} m\, P_m = N$. Equation 69 should apply to the relaxation of stress for linear chains which are attached at one end to a permanent network [12, 32].

2. *Star Polymers*

We assume that disengagement by pure reptation is negligible for star molecules with sufficiently long arms in an entangled medium. (For a contrary opinion, however, based on computer simulation of star molecule motions, see Ref. 33). Relaxation for an f-arm star in a topologically invariant medium is then equivalent to the relaxation of f tethered chains, where τ_e is the tethered chain relaxation time (Eq. 66) for individual arms (Fig. 12).

Constraint release is likely to be very important in the relaxation of branched polymer liquids[14]. However, if we ignore that complication, the stress relaxation modulus for a liquid of highly entangled stars is given simply by Eq. 69 with ν replaced by νf. The viscosity and recoverable compliance can then be calculated from Eq. 69 with Eqs. 25 and 26.

$$\eta_0 = \frac{4\,\tau_e \nu f kT}{5} \sum_{n=1}^{m_0} \frac{Q_n}{-\ln Q_n} \tag{71}$$

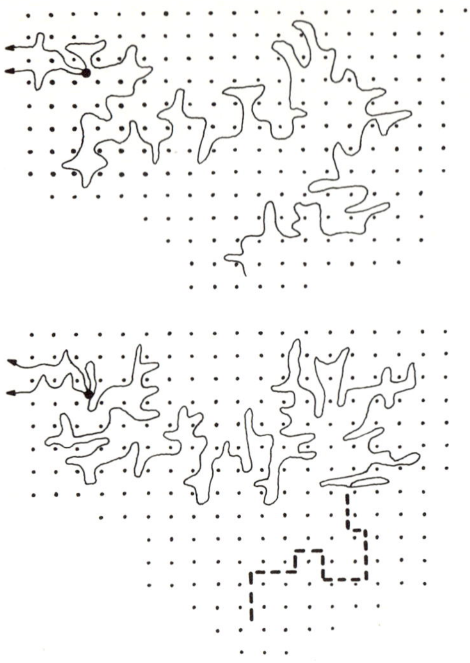

Fig. 12. Abandonment of part of the initial primitive path of one arm of a 3-arm star by breathing. A fluctuation in the surplus segment population results in the abandonment of the dashed portion of the path

$$J_e^0 = \frac{5}{4f\nu kT}\left(\sum_{n=1}^{m_0}\frac{Q_n}{\ln^2 Q_n}\right)\left(\sum_{n=1}^{m_0}\frac{Q_n}{\ln Q_n}\right) \tag{72}$$

Numerical calculations using Eq. 50 for P_m with $p = 0.25$, 0.5 and 0.75 and a wide range of values of $\langle m \rangle = N_b$ (N_b = average number of primitive path steps per arm) are expressed with good accuracy (~10%) by the following equations

$$\eta_0 = 1.1\frac{4\tau_e\nu fkT}{5}\exp\left[\frac{\ln(1/p)}{(1-p)}N_b\right] \tag{73}$$

$$J_e^0 = 0.85\frac{5}{4f}\frac{1}{\nu kT} = \frac{1.06}{f}\frac{M}{cR_GT} \tag{74}$$

With Eqs. 9a and 27 for linear chains and $N = fN_b$, the ratio of viscosities for star and linear polymers of the same total chain length is

$$\frac{(\eta_0)_B}{(\eta_0)_L} = 1.1\frac{8}{f^3\pi^2}\frac{1}{N_b^2}\exp\left[\frac{\ln(1/p)}{(1-p)}N_b\right] \tag{75}$$

where N is the same for both molecules, $N_b = N/f$ is large, and τ_e (arm) $= 4/f^2\,\tau_e$ (entire linear chain) from Eq. 66. Thus, $(\eta_0)_B$ is predicted to grow exponentially with arm length, as observed experimentally[19, 20]. The prefactor in Eq. 7 ($\propto N_b^{-2}$) is different from that proposed by Doi and Kuzuu[9] ($\propto N_b^0$), but the exponential will certainly dominate in any case. The exponential term is predicted to be independent of branch point functionality; experimentally that term increases with f and appears to become independent only at large f[19]. Again, however, it must be noted that both the Doi-Kuzuu predictions and those presented here are based on the assumption of topologically invariant surroundings. Constraint lifetime effects may modify the results in branched polymer liquids considerably (see Chap. III).

The recoverable compliance is predicted to vary quite differently with molecular weight and concentration in entangled linear and star polymers. Thus, from Eq. 28 and data in Ref. 13,

$$J_e^0 \sim (G_N^0)^{-1} \propto c^{-2}M^0 \qquad \text{(linear)} \tag{76}$$

while, from Eq. 74,

$$J_e^0 \propto c^{-1}M^1 \qquad \text{(star)} \tag{77}$$

This latter result is now well established experimentally[34–36]. Even the numerical coefficient in Eq. 74, 1.06/f, is roughly correct, although the smaller "Rouse model" coefficient, $2(15\,f\text{-}14)/5(3\,f\text{-}2)^2$, is in even better agreement.

Both experimentally and theoretically the relaxation spectrum for entangled stars differs quite substantially in form from the Rouse spectrum. Thus, at first sight it seems remarkable that the values of J_e^0 should resemble so closely the Rouse model predictions. However, the reason is quite simple. The "Rouse form" ($J_e^0 \propto c^{-1}M^1$) is expected whenever the intensity of the longest relaxation time (τ_1) is proportional to the concentration of polymer molecules (ν) and the remaining relaxation times are much smaller than τ_1. This is the case both for the Rouse model and for entangled stars (see Appendix IV), so the required summations give $J_e^0 \propto \nu^{-1} \propto c^{-1}M^1$ for each. The relaxation spacings are quite different for the two spectra, but even the numerical coefficient for J_e^0 is rather insensitive to these differences. Thus, it seems unlikely that the predictions about J_e^0 for entangled star liquids will be changed significantly when contraint release is included.

Long branches should severely reduce the center of gravity diffusion coefficient, if simple reptation is suppressed. In a topologically invariant medium only the fluctuations in path length are available to provide long range mobility. Consider again the case of star molecules. Each arm occupies an average of N_b path steps, but at any moment there is some fraction P_0 of arms which occupy no steps (P_0 from Eq. 50 for instance). Each time that event occurs in a 3-arm star the path is topologically linear. The molecule can then reptate until equilibration restores the branch point constraint. The frequency of this occurrence for a 3-arm star is $\phi = 3\,P_0/\tau_e$, and we estimate that the center of gravity moves a distance of about one primitive path step each time ($\overline{\Delta \mathbf{R}g \cdot \Delta \mathbf{R}g} \approx a^2$). From Eq. 18,

$$D = \frac{P_0 a^2}{2\tau_e} \qquad \text{(3-arm star)} \tag{78}$$

or, with Eqs. 1, 4, 21 and 66,

$$D = 9\pi^2 D(2N_b)P_0 \qquad \text{(3-arm star)} \tag{79}$$

where $D(2N_b)$ is the diffusion coefficient in the same medium of a linear chain which occupies $2N_b$ path steps (the transient diffusing species). For f-arm stars a movement can occur only when f-2 arms occupy no steps during the same equilibration interval. The probability of this event is $\frac{f!}{2!(f-2)!} P_0^{f-2}$, so with Eq. 53 (maximum randomness):

$$D = 3\pi^2 \frac{f!}{(f-2)!2!} D(2N_b) \exp\left[-\frac{\ln(1/p)}{1-p}(f-2)N_b\right] \quad \text{(f-arm star)} \tag{80}$$

Thus in a medium of invariant topology the diffusion coefficient of long arm stars should be extremely small, of the order of the diffusion coefficient of the transient (linear) reptating species multiplied by a factor which decreases exponentially with both arm length and the number of arms.

3. Linear Polymers

Path length fluctuations (breathing or tube leakage) will compete with pure reptation in the case of unattached linear chains, although the effect will die out for sufficiently long chains. Suppose all chains occupy N steps initially. For pure reptation the fraction of still surviving steps at position j, measured from the center of the initial path ($j = o, \pm 1, \pm 2, ----, \pm N/2$), is

$$F_j(t) = \frac{4}{\pi} \sum_{\substack{n \\ \text{odd}}} \frac{(-1)^{\frac{n-1}{2}}}{n} \cos \frac{n\pi j}{N} \exp\left[-\frac{n^2 t}{\tau_d}\right] \tag{81}$$

which is obtained from Eq. 7 by replacing x/L with $(j/N + 1/2)$. However, some of these positions will have already been abandoned as a result of path breathing, so $F_j(t)$ is reduced ley some factor $K_j(t)$. If the effects of the two processes within each chain are uncorrelated,

$$G(t) = \frac{4}{5} \nu N k t \left[\frac{2}{N} \sum_{j=1}^{N/2} F_j(t) K_j(t)\right] \tag{82}$$

for large N.

A crude estimate of the breathing effect can be made by treating the linear chain as if it were tethered at its midpoint, i.e., like a “2-arm star“, thus completely ignoring the effect of its diffusion history along the tube. In that case

$$K_j(t) = Q_j^{t/\tau_e} \tag{83}$$

where τ_e corresponds to a tethered chain (Eq. 66) with one half the length of the linear molecule. This will surely underestimate the contribution of tube leakage since any displacement of the chain along its initial tube will move a chain end into the initial tube interior, thereby increasing the likelihood of disengagement by breathing for steps near the middle. On the other hand, it should not be too difficult to program a numerical simulation for the combined process. Doi has recently published an interesting analytical approach for combining reptation and breathing effects in linear chains[10)].

III. Constraint Release

So far we have considered only the contributions of chains which are moving among obstacles of fixed topology. The previous calculations of diffusion coefficient and stress relaxation take no account of rearrangements in the surroundings. The large scale conformation of a chain is defined by the shape of its primitive path among permanent obstacles. Memory of any initial conformation is preserved in the orientations of those initial path steps which are still occupied. That memory is perfect in the sense that the orientations of any surviving steps are precisely determined by the unchanging topology of the surroundings.

For entangled liquids the approximation that motions in the surroundings can be ignored needs to be examined carefully even if the dominant motion is reptation. All chains are moving, and the confining obstacles for any chain are themselves parts of similarly reptating neighbors. The constraints which comprise the tube have finite lifetimes. Each such constraint is provided by a neighboring chain and is released when that chain abandons the particular step along its own primitive path which provides the constraint (Fig. 13). A local displacement of the path can then occur. The effect, over time, is a rearrangement of the path conformation by random local jumps[21)]. Thus, the memory of any chain for past conformations fades away by path rearrangement, produced by disengagement events in its surroundings as well as by its own disengagement actions.

Consider a liquid of linear chains, ignore path breathing, and assume that reptation is the dominant motion. The usual Doi-Edwards approximations apply: an individual chain always occupies the same number of primitive path steps; surplus segments are distributed uniformly along the path. The surroundings will be represented by a regular lattice of independently reptating chains (lattice spacing equal for simplicity to the Doi-Edwards path step length a). A chain with N path steps occupies N cells of the lattice. Each cell is bounded by z_0 "bars" which now are strands of z_0 neighboring chains, occupying their own set of primitive path steps. In a network the bars would be permanent, reflecting the topological permanence of the surroundings. In a liquid they disappear and reform occasionally. Their lifetimes correspond to the lifetimes of primitive path steps which are at random locations along the neighboring chains.

A. Diffusion and Small Strain Response

Constraint release is here represented by a temporary removal of individual cell bars. During the open time the chain can move into the available adjacent cell, thereby altering its path locally. The probability of chain movement will depend in each case on the dispositions of the adjoining portions of its path. We will assume the choice (jump or

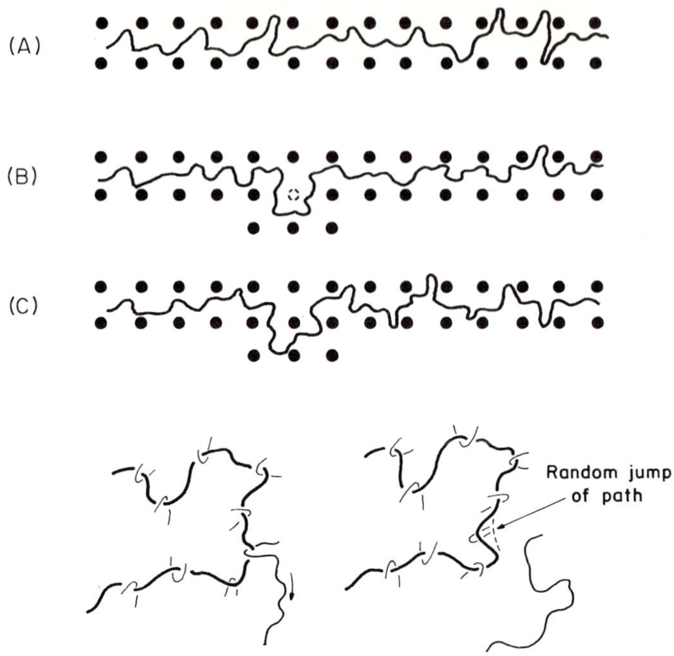

Fig. 13 A–C. Release of a constraint on the primitive path by reptation of a neighboring chain. The initial path (**A**) is altered when the end of a neighboring chain reptates past the path (**B**) and is replaced by a new chain (**C**). The result is a local random jump in conformation (*lower figure*)

no jump) is random when a jump would not alter the path length, but that otherwise no jump occurs. This "length preserving" assumption seems drastic, but it really amounts to nothing more than the omission of fluctuations. From equilibrium considerations (applicable for diffusion in the quiescent liquid or for relaxation from sufficiently small strains), the changes in path length must average to zero, and the probability of a jump must be reduced, on average, for choices which either greatly lengthen or greatly shorten the path.

That assumption simplifies the analysis of primitive path rearrangement. Local path jumps now correspond to random flips in the Orwoll-Stockmayer chain model[37)], and we can apply these results directly. For our case the local jump distance is the path step length a, and the average time between jumps is $2\tau_w$, where τ_w is the mean waiting time for release of a constraint which allows a length preserving jump. The average number of such "suitably situated" constraints per cell is $z(z < z_0)$, and we assume for simplicity that all cells have z such constraints.

For $z = 1$ the mean waiting time is just the average lifetime of primitive path steps in the system. For monodisperse linear chains the fraction of steps which still survive after time t is given by Eq. 8. Thus the average lifetime is

$$\langle t \rangle_1 = \int_0^\infty F(t)\,dt = \frac{\pi^2}{12}\tau_d \tag{84}$$

For $z > 1$ a jump can occur when the first suitably situated constraint is released. The mean waiting time is then

$$\tau_w = \langle t \rangle_z = \int_0^\infty [F(t)]^z dt \tag{85}$$

and with Eq. 8,

$$\tau_w = \Lambda(z)\tau_d \tag{86}$$

where

$$\Lambda(z) = \left(\frac{8}{\pi^2}\right)^z \underbrace{\sum_{n_1} \cdots \sum_{n_z}}_{\substack{\text{odd values} \\ \text{of the indices}}} \frac{1}{n_1^2 + n_2^2 \cdots + n_z^2} \frac{1}{n_1^2 \cdot n_2^2 \cdot \cdots \cdot n_z^2} \tag{87}$$

Values for Λ calculated from Eq. 87 lie between $(8/\pi^2)^z = (.811)^z$ and $(\pi^2/12)^z = (0.822)^z$. Thus, for definiteness and with an error which is insignificant in view of the other simplifications,

$$\Lambda(z) = \left(\frac{\pi^2}{12}\right)^z \tag{88}$$

Equation 85 applies also for a medium of long arm star molecules but with $F(t)$ given by Eq. 68. In polydisperse systems the fraction of path steps contributed by any species is equal to its weight fraction, and presumably some average based on Eq. 85 would apply. The following discussion of diffusion and stress relaxation will consider only situations which involve monodisperse surroundings.

The center of gravity of a linear chain now moves by two uncorrelated processes, reptation and constraint release, so the diffusion coefficient is just the sum of the individual contributions. Equation 21 gives the reptation contribution. Equation 18 gives the constraint release contribution with $\phi = N/2\tau_w$ and $\overline{(\Delta R_g)^2} = a^2/N^2$ for chains which occupy N primitive path steps:

$$D = \frac{1}{12} \frac{a^2}{N\tau_w} \quad \text{(constraint release contribution only)} \tag{89}$$

The waiting time τ_w depends on the nature of the surroundings. If the diffusing species is a dilute component in linear chains which occupy N_s steps, the constraint release contribution (with Eqs. 2, 9 and 86) is

$$D(N, N_s) = \frac{\pi^2}{12} \frac{1}{\Lambda N_s^3} D^*(N) \tag{90}$$

Thus, the diffusion coefficient including both processes is

$$D(N, N_s) = \left[\frac{1}{3N} + \frac{\pi^2}{12\Lambda N_s^3}\right] D^*(N) \tag{91}$$

The ratio of reptation contribution to constraint release contribution is thus $4\Lambda N_s^3/\pi^2 N$, so, with $\Lambda(z)$ fixed, the reptation contribution dominates for sufficiently large N in the case of self diffusion ($N_s = N$). Reptation should always dominate, and D should be independent of the matrix, when the diffusing species is smaller than the matrix ($N < N_s$). The diffusion coefficient should vary as $N^{-1}N_s^{-3}$ (or $M^{-1}M_s^{-3}$) when the diffusing species is sufficiently large compared to the matrix. Variations with N_s in this case could be used, at least in principle, to estimate a value for the constraint parameter $\Lambda(z)$. However, Eq. 91 merely reaffirms earlier theoretical conclusions[21, 38)], supported by recent experimental results[39, 40)], that diffusion of linear chains is dominated by reptation and insensitive to matrix chain length. Equation 91 also omits the effects of intramolecular hydrodynamic interactions, which is probably not correct when $N \gg N_s$[41)].

The situation should be quite different for star molecules however. Equation 90 gives the constraint release contribution for dilute stars in a linear chain matrix, and Eq. 80 gives the reptation contribution. The exponential term makes the latter very small for long arm stars, so one would expect a strong matrix dependence ($D \propto N_s^{-3}$), providing a rather direct way to estimate $\Lambda(z)$. Equation 89 should apply also for a star matrix, but with τ_w calculated from Eq. 85 with F(t) given by Eq. 68.

The contributions of path displacement to linear viscoelastic properties can be obtained using the bond flip model[37)]. The stress relaxation modulus for that model is

$$G(t) = \nu kT \sum_{j=1}^{N} \exp\left[-\alpha\lambda_j(N)t\right] \tag{92}$$

where ν is the number of chains per unit volume, N is the number of "bonds" (primitive path steps) per chain, α is a constant which depends on jump distance and bond length, and the λ_j are the Rouse chain eigenvalues:

$$\lambda_j(N) = 4\sin^2\left[\frac{\pi j}{2(N+1)}\right] \qquad j = 1, 2, ----, N \tag{93}$$

If constraint release were the only process for conformational rearrangement, the initial path motions would be the same as the chain motions of the N-element Rouse model. Equation 4 relates the longest relaxation time τ_1 to the diffusion coefficient for Rouse chains. The diffusion coefficient from constraint release is given by Eq. 90 with $N_s = N$. With Eqs. 1, 2, 4 and 9,

$$\tau_1 = \frac{2\Lambda}{\pi^2} N^2 \tau_d \tag{94}$$

and, since the longest relaxation time is $(\alpha\lambda_1(N))^{-1}$,

$$\alpha = \frac{1}{2\Lambda\tau_d} \tag{95}$$

from Eq. 93 for N large. Thus, for chains which relax only by release of tube constraints,

$$G_p(t) = \nu kT \sum_{j=1}^{N} \exp\left[-\frac{\lambda_j(N)t}{2\Lambda\tau_d}\right] \tag{96}$$

The effects of simultaneous reptation and constraint release can be estimated if we assume the processes act independently. At time t following a small shear strain γ (and after the initial equilibration) the rate of stress relaxation, $-d\sigma/dt$, is the sum of two terms:

(i) the rate due to constraint release per initial path step, $-\frac{4}{5}\frac{\gamma}{\nu N}\frac{dG_p}{dt}$, multiplied by the concentration of initial path steps which are still occupied, $\nu N F(t)$.

(ii) the rate due to abandonment of initial path steps, $-\nu N\frac{dF}{dt}$, muiltiplied by the stress contribution per initial path step which is still occupied, $\frac{4}{5}\frac{\gamma}{\nu N}G_p(t)$.

Thus,

$$\frac{5}{4\gamma}\frac{d\sigma}{dt} = \frac{dG_p}{dt}F(t) + \frac{dF}{dt}G_p(t) \tag{97}$$

which integrates to give

$$G(t) = \frac{\sigma(t)}{\gamma} = \frac{4}{5}\nu NkT\,F(t)\,R(t) \tag{98}$$

where

$$R(t) = \frac{1}{N}\sum_{j=1}^{N} \exp\left[-\frac{\gamma_j(N)t}{2\Lambda\tau_d}\right] \tag{99}$$

and F(t) is given by Eq. 8.

Expressions for viscosity and recoverable compliance are readily obtained with Eqs. 25 and 26:

$$\eta_0(\Lambda, N) = \frac{G_N^0\tau_d}{N}\sum_{i\ \mathrm{odd}}\sum_{j=1}^{N}\frac{1}{i^2}\left(\frac{2\Lambda}{2\Lambda i^2 + \lambda_j(N)}\right) \tag{100}$$

$$J_e^0(\Lambda, N) = \frac{N}{G_N^0}\frac{\left[\sum_{i\ \mathrm{odd}}\sum_{j=1}^{N}\frac{1}{i^2}\left(\frac{2\Lambda}{2\Lambda i^2 + \lambda_j(N)}\right)^2\right]}{\left[\sum_{i\ \mathrm{odd}}\sum_{j=1}^{N}\frac{1}{i^2}\left(\frac{2\Lambda}{2\Lambda i^2 + \lambda_j(N)}\right)\right]^2} \tag{101}$$

The quantities $\eta_0/G_N^0\tau_d$ and $J_e^0G_N^0$ become functions of Λ alone at large N ($N \gtrsim 20$). Table 1 shows the limiting values for some choices of z.

Table 1.

z	Λ(Eq. 88)	$\eta_0/G_N^0\tau_d$	$J_e^0G_N^0$
0^a	∞	0.822	1.20
1	0.822	0.417	1.50
3	0.556	0.361	1.62
6	0.310	0.298	1.86
10	0.142	0.216	2.39

[a] Doi-Edwards values (Eqs. 27 and 28) for pure reptation

The limiting laws predicted by the Doi-Edwards model, $\eta_0 \propto M^3$ and $J_e^0 \propto M^0$, are thus not altered by constraint release. However,in contrast to the diffusion coefficient (Eq. 91), the numerical values of η_0 and J_e^0 for large N can be changed rather substantially by constraint release. The viscosity is reduced by a factor of $2 \sim 3$ for z in a physically plausible range, $3 \sim 6$. Furthermore, this would translate to a roughly similar reduction in longest relaxation time for linear chains in the homologous melt (where reptation, breathing and constraint release would all contribute) relative to the unattached chains in a network (where only reptation and breathing would contribute). Experimentally, this reduction factor is about 3 and independent of chain length for linear polymers[14]. The $J_e^0G_N^0$ product also moves up toward the experimental value ($J_e^0G_N^0 \approx 3$) when constraint release is included. The values for $z = 3 \sim 6$ are still below the experimental result, but that result itself is undoubtedly larger than the value that would be obtained in truly monodisperse systems[12, 13]. Equations for calculating linear viscoelastic properties in the limit of large N, where path breathing effects should become negligible, are given in Appendix IV.

Differences in sensitivity to constraint release between η_0 and J_e^0 on the one hand and D on the other come from differences in their weighting of the local motions. Center of gravity diffusion depends only on the largest scale chain motions, and these are practically unchanged by constraint release as long as τ_1 (Eq. 94) $\gg \tau_d$. The values of η_0 and J_e^0 depend on the pre-exponential intensity factors as well. The shortest relaxation times from constraint release ($j \approx N$ in Eq. 99) remain quite comparable to the disengagement time even at large N. Thus, for example, $\tau_N = \Lambda\tau_d/2$. These serve, in effect, to reduce the intensity factors of the reptation contribution, thus decreasing the viscosity and increasing the recoverable compliance for monodisperse linear polymers. Arguments which dismiss constraint release based simply on comparisons of the longest relaxation times with those from reptation must therefore be used with some caution.

Equation 98 is also applicable to monodisperse star polymers except in this case F(t) is given by Eq. 68, R(t) requires the eigenvalues appropriate to the star structure in Eq. 99, and $\Lambda\tau_d$ must be replaced by τ_w calculated from Eq. 85. Values of τ_w can be estimated by using the properties of P_m for small m when N_b is large (see Appendix IV). In this case,

$$\tau_w = S_3 \frac{\tau_e}{zN_b^zP_0} \tag{102}$$

where S_3, a numerical factor of order unity, the sum of a series which is insensitive to N_b, z and the detailed form of P_m. (For example, $S_3 = 1.16$, 1.28 and 1.59 are obtained for z = 1, 3 and 6 using Eq. 50 with p = 0.5.) The longest relaxation times in F(t) for stars are τ_e/P_0, τ_e/N_bP_0, $2\tau_e/N_b^2P_0$, etc. (Appendix IV with P_m values calculated from Eq. 53), and the largest eigenvalues in R(t) are approximately 4. Thus, the gross effect of constraint release in stars is to shift all contributions in F(t) with relaxation times larger than τ_w down to τ_w, and to leave the faster processes largely unaffected. As a result, terminal relaxation time for homologous star melts should be smaller than that for unattached stars in a network by a factor of roughly zN_b^z. Assuming a value of $z \sim 3$, this result at least qualitatively consistent with the extremely large and structure dependent shifts which have been observed for star polymers in the two environments[14].

Relaxation of dilute species (star or linear) in monodisperse matrices (star or linear) can also be worked out with Eq. 98, using F(t) appropriate to the dilute species and R(t) for the matrix. Preliminary results indicate that long-arm molecule relaxations can be controlled over wide ranges by the choice of chain length for a linear polymer matrix. On the other hand, relaxations of linear chains in a star matrix should be less affected by matrix chain structure. Their behavior should move from the homologous linear melt behavior when τ_w in the matrix is of the order of τ_d for the linear chains to behavior as unattached linear chains in a network when $\tau_w \gg \tau_d$. The latter prediction seems inconsistent with recent experimental results[42] where linear chain relaxation rates were found to be the same in the homologous melt and in a star matrix with $\tau_w \gg \tau_d$. This may indicate some fundamental problem with the equation suggeted to estimate τ_w (Eq. 85).

Application of these ideas to general polydisperse systems will require the extension of mean waiting time calculations to mixed structure matrices.

B. Constraint Release for Large Strains

Two additional factors related to constraint release need to be considered for finite step strain experiments in liquids:

(i) During the initial equilibration following deformation the chains retract in their tubes to re-establish the equilibrium chain density per unit path length[3]. In the Doi-Edwards picture, the number of path steps changes from N to $N/\langle|\mathbf{E}\cdot\mathbf{u}|\rangle$. We have associated each step with a constraint on the path of z_0 neighboring chains. Thus, the retraction process simultaneously releases $z_0\nu N(1-\langle|\mathbf{E}\cdot\mathbf{u}|\rangle^{-1})$ neighboring chain constraints per unit volume, or an average of $zN(1-\langle|\mathbf{E}\cdot\mathbf{u}|\rangle^{-1})$ effective constraints per chain.

(ii) Effects of the subsequent constraint release by reptation in the surroundings will probably be changed when the strain is large. It is likely that the jumps will not obey the step length conserving assumption since, following a finite strain and equilibration, the step lengths of the paths are unequal and the angles between successive steps are not random. Also, the number of path steps will change during the subsequent relaxation: constraint release will itself tend to restore N to its equilibrium value. Thus, the bond flip model (Eq. 96) will become increasingly inapplicable for estimating the path displacement contribution as the strains become very large. Departures from separability of the strain and time dependences can then be expected.

Effects of the first factor, release of constraints by retraction during the equilibration period, can be estimated with the bond flip model as long as the strain is not too large. The average frequency of retraction-induced jumps per chain during equilibration is

$$\phi = \frac{zN}{2\tau_e}\left(1 - \frac{1}{\langle|\mathbf{E}\cdot\mathbf{u}|\rangle}\right) \tag{103}$$

and the center of gravity displacement from each is a/N. The "diffusion coefficient" for the chain during this period is therefore (Eq. 18)

$$D = \frac{1}{12}\frac{a^2 z}{\tau_e N}\left(1 - \frac{1}{\langle|\mathbf{E}\cdot\mathbf{u}|\rangle}\right) \tag{104}$$

and, by the method used to obtain Eqs. 94 and 95,

$$\alpha = \frac{z}{2\tau_e}\left(1 - \frac{1}{\langle|\mathbf{E}\cdot\mathbf{u}|\rangle}\right) \tag{105}$$

Thus, the fraction of stress remaining at the end of the equilibration period ($t \sim \tau_e$) is (from Eq. 96):

$$S(\mathbf{E}) = \frac{1}{N}\sum_{j=1}^{N}\exp\left[-\frac{z\lambda_j(N)}{2}\left(1 - \frac{1}{\langle|\mathbf{E}\cdot\mathbf{u}|\rangle}\right)\right] \tag{106}$$

The stress following equilibration is therefore the product of this term and the stress for chains in an environment of permanent obstacles (Eq. 64):

$$\sigma_{\alpha\beta}\;(\tau_e) = \frac{3\nu NkT}{\langle|\mathbf{E}\cdot\mathbf{u}|\rangle}\left\langle\frac{(\mathbf{E}\cdot\mathbf{u})_\alpha(\mathbf{E}\cdot\mathbf{u})_\beta}{|\mathbf{E}\cdot\mathbf{u}|}\right\rangle S(\mathbf{E}) \tag{107}$$

For small strains $S(\mathbf{E})$ is only negligibly different from unity, so inclusion of this term does not change the expression for plateau modulus (Eq. 37). It also does not affect the factorization of strain and time dependence for $t > \tau_e$, but it does change the strain dependent part rather substantially, principally by shifting the onset of strain dependence in $h(\gamma)$ to smaller strains (Fig. 14).

IV. Summary and Discussion

Part I summarizes the main ideas of de Gennes, Doi and Edwards about tube models and reptation in entangled polymer systems. Attention has been limited to properties for which predictions can be made without invoking the independent alignment approximation: macromolecular diffusion, linear viscoelasticity in the plateau and terminal regions, stress relaxation following a step strain from rest of arbitrary magnitude, and equilibrium elasticity in networks.

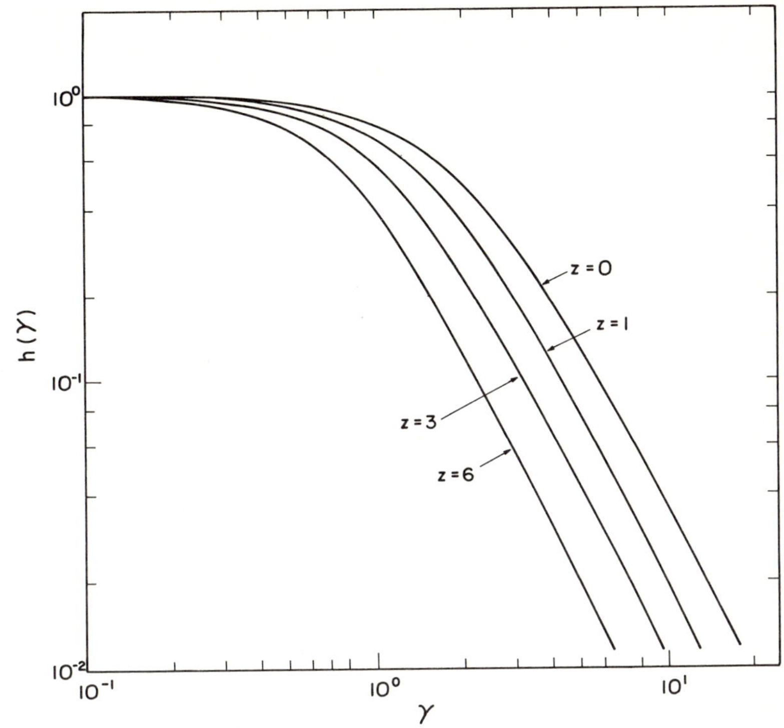

Fig. 14. The strain dependent part of the shear stress relaxation function for different values of the tube constraint parameter z

The Doi-Edwards theory treats monodisperse linear chain liquids by a model which suppresses fluctuations and assumes a topologically invariant medium. Two parameters are required, the monomeric friction coefficient ζ_0 which characterizes the local dynamics and the primitive path step length a which characterizes the topology of the medium. The step length is related to the entanglement molecular weight of earlier theories, $M_e = cR_GT/G_N^0$, by Eqs. 1 and 37:

$$a^2 = \frac{4}{5}\frac{\langle R^2\rangle}{M}M_e \tag{108}$$

The monomeric friction coefficient sets the time scale of all motions, e.g., the disengagement time:

$$\tau_d = \frac{5}{4\pi^2kT}\frac{\langle R^2\rangle}{M}\frac{M^3}{M_0M_e}\zeta_0 \tag{109}$$

when M_0 is the molecular weight of the monomer unit. (Equation 109 is equivalent to Eq. 57a on p. 250 of Ref. 12 when the right hand side of that equation is multiplied by 5/4.) However, the Doi-Edwards theory provides much more than a mere translation of terminology. It yields relationships between properties, G_N^0 vs J_e^0 and G_N^0 vs D, which agree remarkably well with observations on linear polymer liquids. It also fails in several ways, but even its failures are illuminating and suggestive of competing relaxation mechanisms which may even supplant reptation in some situations.

Part II proposes a new model which still assumes a topologically invariant medium but provides for fluctuations in the primitive path length (tube leakage or path breathing) as well as simple reptation. This introduced a third parameter, p, which characterizes the distribution of path lengths for a chain. The relaxation of tethered and branched chains is considered as well as the competition between reptation and tube leakage in the relaxation of linear chains. The elasticity equation obtained for entangled networks is in better agreement with observations than that from the Doi-Edwards model. The Doi-Edwards equations are recovered for sufficiently long linear chain liquids, with

$$a = \frac{d}{(1-p)^{1/2}} \tag{110}$$

where d is the mesh size, the topological parameter in the proposed model.

Part III deals with constraint release, the relaxation of chains when the topology of the medium has a finite lifetime. Constraint release, operating in liquids but not in networks, is related to the average lifetime of path steps in the system. The influence of constraint release on changes in the path conformation is governed by a fourth parameter z, an effective coordination number for the constraint lattice. Differences in the relaxation rates of linear and branched chains in liquids and in networks are considered. Equations are developed for predicting the combined effects of reptation and constraint release in monodisperse linear chain liquids and in linear and branched chain mixtures.

The values of all four parameters – ζ_0, a (or d), p and z – should be independent of the large scale structure of the chains, but they vary from one system to another in different ways.

The monomeric friction coefficient has been studied extensively[12]. As in small molecule dynamics, ζ_0 depends on many variables (polymer and solvent species, concentration, temperature, etc.), but its independence of chain length beyond a certain length is well established. The value of ζ_0 can be estimated by a variety of methods[12]. The path step length a is virtually a universal function of the total length of chain backbone per unit volume in the system[43], and quite insensitive to all other variables. Furthermore, for a given polymer species, $a \propto \phi^{-0.5 \sim 0.6}$, where ϕ is the volume fraction of polymer. The latter is based on the observation that $G_N^0 \propto \phi^{2.0 \sim 2.2}$ [42], and $G_N^0 \propto \phi/a^2$ from Eqs. 1, 2 and 37.

The properties of the surplus segment probability p and the effective constraint coordination number z are less well established. It seems possible that p will depend on polymer species to some extent, since loop projection may be easier for a more locally flexible chain. Weak dependences on concentration and temperature are likely for the same reason. On the other hand, z characterizes the topology on a fairly large scale and therefore may be virtually a universal constant. These however are only some speculations. Values of p and z can be established by various experiments, p from the elastic properties of networks and also from the relaxation of star polymers, z from relative relaxation rates of linear and star molecules in liquids and networks and also from measurements of diffusion rates of stars in linear chain liquids. The adequacy of the equations presented here for the path length distribution P_m and the mean waiting time τ_w must be judged by the agreement between values of p and of z obtained from independent experiments.

The aim is finally, with ζ_0, a (or d), p and z alone, to obtain a uniform and self consistent description of all the following properties of entangled systems:

i. Equilibrium elasticity of entangled networks.

ii. The laws relating D, η_0, J_e^0, G_N^0 to M in monodisperse linear chain and star chain liquids.

iii. Relaxation behavior of unattached linear and star molecules in networks.

iv. Relaxation behavior of dilute linear and star molecules in linear and star entangled matrices.

v. Polydispersity effects over the entire range in binary mixtures (linear-linear, star-star and linear-star).

Future publications will present more detailed comparisons with experimental data.

Two theories of viscoelasticity in reptating chain systems have appeared since the original Doi-Edwards publications, the theory of Marrucci and coworkers[29] and the theory of Curtiss and Bird[30]. They differ in various ways from the Doi-Edwards tube model and from the model suggested in Part II. It is difficult and probably premature to provide detailed criticisms at the present time, but it is perhaps worthwhile to point out at least a few of the differences.

Like Doi and Edwards, Marrucci and Hermans consider the properties of a chain confined to the vicinity of a random walk path by a quadratic potential. Deformation transforms the path affinely, and it is supposed that the spatial parameters of the potential field, representing the surrounding constraints and defining the effective tube diameter, also change affinely. Doi and Edwards consider that possibility, but go on to treat the potential as independent of deformation, in effect assuming that the tube diameter and the chain density per unit path length (for $t > \tau_e$ in a step strain) is not changed by deformation. The effect of tube diameter changes vanishes in the limit of linear viscoelasticity, so in those properties the theories agree. The differences in non-linear liquid response and in network elasticity can apparently be substantial, however.

An important question is whether a potential field is a proper and physically reasonable way to handle the constraint effects in a deformed system. The surroundings are, after all, a set of uncrossable strands, and the density of these does not change with deformation. There is no reason to suppose that the strands move closer together in some directions and further apart in others, as would seem to be required by an affine displacement of constraints. On the other hand, the assumption of constant constraint spacing also presents some conceptual problems in the Doi-Edwards model. In calculating the stress, Doi and Edwards assume that the segment of chain in each primitive path step acts as a Gaussian chain with end-to-end vector parallel to the step vector **u** but otherwise able to explore all conformations freely. For free exploration, the equilibrium tube diameter must be of the order of the step length a, since $\langle r^2 \rangle \sim a^2$ for its chain segment. After deformation **E** the step length changes to $a|\mathbf{E} \cdot \mathbf{u}|$; after equilibration the chain density along the path returns to its equilibrium value, so the amount of chain along that step changes by the factor $|\mathbf{E} \cdot \mathbf{u}|$. But now the tube diameter must change by the factor $|\mathbf{E} \cdot \mathbf{u}|^{1/2}$ in order for free exploration of conformation to still apply: if a step is stretched, its tube diameter must also increase. This is opposite to Marrucci's affine tube diameter assumption (conservation of tube "volume") and inconsistent with the constant diameter assumption of Doi-Edwards. The model presented in Pt. II does not share these inconsistencies. The tube diameter never enters either explicitly or implicitly. The stress comes from the orientation of primitive segments whose size depends only on the obstacle spacing. This produces the difference from Doi-Edwards in the equilibrium elasticity of entangled networks, although that difference vanishes in the stress law (after equilibration) for entangled liquids.

Curtiss and Bird introduce reptation in a manner which does not involve the tube concept, at least not in an explicit way. Their model leads to a constitutive equation in which the stress is the sum of two contributions. On contribution is exactly 1/3 the expression obtained by Doi and Edwards when those workers invoke the independent alignment approximation, i.e., that contribution is a special case of the BKZ relative strain equation. The other contribution depends on strain rate and is proportional to a "link tension coefficent" ε $(0 \le \varepsilon \le 1)$ which characterizes the forces along the chain arising from the continued displacements of chain relative to surroundings.

Even aside from the rather puzzling factor of 1/3 difference, the Curtiss-Bird result does not reduce to Doi-Edwards in the limit of linear viscoelasticity. For example, in Curtiss-Bird the dynamic viscosity $\eta'(\omega) \equiv G''(\omega)/\omega$ changes from η_0 at $\omega = 0$ to $\frac{2/3\,\varepsilon}{1 + 2/3\,\varepsilon}\,\eta_0$ at $\omega = \infty$. Experimentally, $\eta'(\omega)$ decreases from η_0 by at least two or three orders of magnitude at high frequencies in concentrated long chain liquids (see, for example, Fig. 8.1 in Ref. 13). In the context of the model, ε must than be chosen very small indeed $(< 10^{-2} \sim 10^{-3})$ to avoid serious inconsistency with the observed linear viscoelastic behavior. Therefore (using values of ε in this range) the strain-rate dependent part of the Curtiss-Bird model must be negligible except for flows with extremely large strain rates.

Acknowledgements. This work was supported by the Science Research Council during 1979–80 while the author was Senior Visiting Fellow at Cambridge University and by the National Science Foundation (CPE-8000030) since that time. The generous encouragement of Professor S. F. Edwards, discussions with Robin Ball, Ken Evans and Nigel Goldenfeld at the Cavendish Laboratory on various parts of the work, computations performed there by Chris Nex, and numerous helpful suggestions on the manuscript by Professor J. D. Ferry and Dr. D. S. Pearson are acknowledged with gratitude.

Appendix I

Properties of P_m When Surplus Segments are Locally Paired

Let 1-p be the probability that a primitive segment selected at random is part of the path-occupying population. If m of a total of m_0 segments are path-occupying segments, there are m + 1 sites over which to distribute the $m_0 - m$ surplus segments. These must occur in pairs to achieve self-cancelling of local loops. The probability of finding a chain with $(m_0 - m)/2$ such self-cancelling pairs and any specified distribution of the pairs among the m + 1 sites is proportional to $(1-p)^m(p)^{\frac{m_0-m}{2}}$. The number of ways to distribute $(m_0 - m)/2$ pairs among m + 1 sites is $\left(\frac{m_0+m}{2}\right)!/m!\left(\frac{m_0-m}{2}\right)!$. The product of these two terms, after normalization, gives

$$P_m = \frac{\left(\frac{m_0+m}{2}\right)!}{m!\left(\frac{m_0-m}{2}\right)!}\ \frac{p^{\frac{m_0-m}{2}}(1-p)^m(1+p)}{\left(1+p^{m_0+1}\right)} \tag{I-1}$$

when $m_0 - m$ is even; $P_m = 0$ when $m_0 - m$ is odd.
When m_0 is large,

$$\langle m\rangle = \frac{1-p}{1+p}m_0 \tag{I-2 a}$$

$$\langle m^2\rangle - \langle m\rangle^2 = \frac{4p}{(1+p)^2}\langle m\rangle \tag{I-2 b}$$

The Gaussian approximation should be valid near the mean for any distribution function when m_0 is large:

$$P(m)dm = \left[\frac{1}{2\pi(\langle m^2\rangle - \langle m\rangle^2)}\right]^{1/2} \exp\left[-\frac{(m-\langle m\rangle)^2}{2(\langle m^2\rangle - \langle m\rangle^2)}\right] dm \tag{I-3}$$

For this distribution, From Eq. I-2,

$$P(m)dm = \left[\frac{(1+p)^2}{8p\langle m\rangle}\right]^{1/2} \exp\left[-\frac{(m-\langle m\rangle)^2(1+p)^2}{8p\langle m\rangle}\right] dm \tag{I-4}$$

When m_0 is large and m is small, Eq. I-1 gives

$$P_m = \frac{(1+p)}{m!}\left[\frac{(1+p)\langle m\rangle}{p^{1/2}}\right]^m \exp\left[-\frac{(1+p)\ln(1/p)}{2(1-p)}\langle m\rangle\right] \tag{I-5}$$

Thus, results obtained with the binominal distribution (Eq. 50) in the Gaussian approximation (Eq. 52) are the same as those with the paired distribution (Eq. I-1) with p replaced by $4p/(1+p)^2$ (Eq. I-4). Likewise, results dominated by the contribution from small m terms will be the same when $\ln(1/p)/(1-p)$ (Eq. 53) is replaced by $(1+p)\ln(1/p)/2(1-p)$ (Eq. I-5).

Appendix II

Entanglement Contribution to Equilibrium Elasticity of Networks

Equations 62 and 63 can be combined to give

$$W = \frac{15}{4} G_N^0 \left[I - 1 + \frac{1}{6p} (I-1)^2 \right] \tag{II-1}$$

where

$$I = \langle |\mathbf{E} \cdot \mathbf{u}| \rangle \tag{II-2}$$

For uniaxial extension

$$\mathbf{E} = \begin{vmatrix} \lambda & 0 & 0 \\ 0 & \lambda^{-1/2} & 0 \\ 0 & 0 & \lambda^{-1/2} \end{vmatrix} \tag{II-3}$$

where λ is the extension ratio ($\lambda \geq 1$; $\lambda \leq 1$ corresponds to compression), and

$$I(\lambda) = \frac{1}{2} \int_0^\pi \left[\lambda^2 \cos^2\theta + \frac{1}{\lambda} \sin^2\theta \right] \sin\theta d\theta \tag{II-4}$$

The nominal tensile stress (tensile force/initial area) is $\sigma = dW/d\lambda$, so

$$\sigma = \frac{15}{4} G_N^0 \left[1 + \frac{I-1}{3p} \right] \frac{dI}{d\lambda} \tag{II-5}$$

The following are obtained from Eq. II-4:
<u>$\lambda \geq 1$ tension)</u>

$$I = \frac{1}{2} \left[\lambda + \frac{1}{\lambda^{1/2}(\lambda^3 - 1)^{1/2}} \ln\left[(\lambda^3-1)^{1/2} + \lambda^{3/2}\right] \right] \tag{II-6}$$

$$\frac{dI}{d\lambda} = \frac{1}{2} \left[1 + \frac{3/2}{\lambda^3 - 1} - \frac{1/2(4\lambda^3 - 1)}{\lambda^{3/2}(\lambda^3 - 1)^{3/2}} \ln\left[(\lambda^3-1)^{3/2} + \lambda^{3/2}\right] \right] \tag{II-7}$$

$\lambda \leq 1$ (compression)

$$I = \frac{1}{2}\left[\lambda + \frac{1}{\lambda^{1/2}(1-\lambda^3)^{1/2}} \tan^{-1}\left(\frac{1-\lambda^3)^{1/2}}{\lambda^{3/2}}\right)\right] \tag{II-8}$$

$$\frac{dI}{d\lambda} = \frac{1}{2}\left[1 - \frac{3/2}{1-\lambda^3} + \frac{1/2(4\lambda^3-1)}{\lambda^{3/2}(1-\lambda^3)^{3/2}} \tan^{-1}\left(\frac{(1-\lambda^3)^{1/2}}{\lambda^{3/2}}\right)\right] \tag{II-9}$$

The stored energy function for the Doi-Edwards model is (Eq. 40):

$$W = \frac{15}{8} G_N^0 [I^2 - 1] \tag{II-10}$$

so the nominal stress for uniaxial extension is

$$\sigma = \frac{15}{4} G_N^0 I \frac{dI}{d\lambda} \tag{II-11}$$

The nominal stress for a network in which entanglements act like permanent junctions which are displaced affinely (from Eq. 41) is

$$\sigma = \frac{5}{4} G_N^0 (\lambda - \lambda^{-2}) \tag{II-12}$$

Stress-strain curves for the various models are plotted in the Mooney-Rivlin fashion in Fig. 9.

Appendix III

Relaxation Time of a Tethered Rouse Chain

The longest relaxation time for a Rouse chain of length L_0 can be written

$$\tau_1 = \frac{\zeta \ell}{6kT} \frac{1}{\lambda_1} \tag{III-1}$$

where ζ is the friction coefficient per unit length of chain, ℓ is the Kuhn step length and λ_1 is the smallest non-zero eigenvalue in the solution of

$$\frac{d^2x}{ds^2} + \lambda x = 0 \tag{III-2}$$

with the appropriate boundary conditions. For a free chain these conditions are

$$\frac{dx}{ds} = 0 \quad \text{at } s = 0 \text{ and at } s = L_0 \tag{III-3}$$

for which the eigenvalues are solutions of $\sin \lambda^{1/2} L_0 = 0$ or

$$\lambda_j = \left(\frac{\pi j}{L_0}\right)^2 \qquad j = 0, 1, 2 \text{---} \tag{III-4}$$

and $\lambda_1 = (\pi/L_0)^2$. For a chain with one end (s = 0) fixed, the boundary conditions are

$$x(s) = 0 \quad \text{at } s = 0$$

$$\frac{dx}{ds} = 0 \quad \text{at } s = L_0 \tag{III-5}$$

In this case the eigenvalues are solutions of $\cos \lambda^{1/2} L_0 = 0$ or

$$\lambda_j = \left(\frac{\pi_j}{2 L_0}\right)^2 \qquad j = 1, 3, 5 \text{---} \tag{III-6}$$

and $\lambda_1 = \frac{1}{4}\left(\frac{\pi}{L_0}\right)^2$. Thus, from Eq. III-1, the longest relaxation time for a Rouse chain tethered at one end is four times larger than that for a free Rouse chain of the same length.

Appendix IV

Expressions for Viscosity and Recoverable Compliance of Entangled Star Polymers

From Eq. 69, the relaxation times for tethered chains in topologically invariant surroundings are

$$\tau_n = \tau_e/(-\ln Q_n) \qquad n = 1, 2, \ldots, m_0 \tag{IV-1}$$

where τ_e is the equilibration time for a tethered chain (Eq. 66). The cumulative distribution Q_n (Eq. 67) remains near unity for small n, decreases rapidly near $n \sim \langle m \rangle$, and approaches zero for $m \sim m_0$. Its values are $Q_1 = 1\text{-}P_0$, $Q_2 = 1\text{-}P_0\text{-}P_1$, $Q_3 = 1\text{-}P_0\text{-}P_1\text{-}P_2$, etc., so when P_0, P_1, etc. are small (large m_0 and $\langle m \rangle$),

$$\tau_1 = \tau_e/P_0$$

$$\tau_2 = \tau_e/(P_0 + P_1) \tag{IV-2}$$

$$\tau_3 = \tau_e/(P_0 + P_1 + P_2) \text{ etc.}$$

Since also $P_0 \ll P_1 \ll P_2$ etc. (see Eq. 53),

$$\tau_1 = \tau_e/P_0$$
$$\tau_2 = \tau_e/P_1 \quad \text{(IV-3)}$$
$$\tau_3 = \tau_e/P_2 \text{ etc.}$$

where $\tau_1 \gg \tau_2 \gg \tau_3$ etc. Thus, each summation required to evaluate η_0 and J_e^0 (Eqs. 71 and 72) is dominated by its leading term. These expressions can be written

$$\sum_{n=1}^{m_0} \frac{Q_n}{-\ln Q_n} = S_1/P_0 \quad \text{(IV-4)}$$

$$\sum_{n=1}^{m_0} \frac{Q_n}{(-\ln Q_n)^2} = S_2/P_0^2 \quad \text{(IV-5)}$$

where S_1 and S_2 are numerical quantities which are sums with leading terms of unity and succeeding terms diminishing very rapidly with increasing n. The expressions for η_0 and J_e^0 then become

$$\eta_0 = S_1 \frac{4\tau_e \nu fkT}{5P_0} \quad \text{(IV-6)}$$

$$J_e^0 = \frac{S_2}{S_1^2} \frac{5}{4\nu fkT} \quad \text{(IV-7)}$$

in which the particular choice of P_m only affects the expression for P_0 and the values of the two summations.

Insensitivity of S_1 and S_2 to N_b for large N_b is thus to be expected, due simply to the wide spacing of the longest relaxation times. The "Rouse form" ($J_e^0 \propto c^{-1}M^1$) should be obtained for any model in which the intensity of the longest relaxation time is proportional to the concentration of polymer molecules (ν) and in which the other relaxation times are much smaller.

Appendix V

Linear Viscoelastic Properties for Long Monodisperse Chains with Both Reptation and Constraint Release Contributions

With Eqs. 8, 98 and 99 the stress relaxation modulus for linear chains, including both reptation and constraint release, can be written

$$G(t) = \frac{G_N^0}{N} \frac{8}{\pi^2} \sum_{\substack{\text{odd}\\ k}} \sum_{j=1}^{N} \frac{1}{k^2} \exp\left[-\left(k^2 + \frac{\lambda_j}{2\Lambda}\right)\frac{t}{\tau_d}\right] \quad \text{(V-1)}$$

The complex dynamic modulus is obtained from G(t) by

$$G^*(\omega) = G'(\omega) + G''(\omega) = i\omega \int_0^\infty G(t)e^{-i\omega}dt \tag{V-2}$$

with the result

$$G^*(\omega) = \frac{G_N^0}{N}\frac{8}{\pi^2}\sum_{\substack{odd\\k}}\sum_{j=1}^{N}\frac{1}{k^2}\frac{i\omega\tau_d}{\beta_{jk} + i\omega\tau} \tag{V-3}$$

where

$$\beta_{jk} = k^2 + \frac{\lambda_j}{2\Lambda} \tag{V-4}$$

Eq. 93 can be rewritten as

$$\lambda_j = \left[2 \cos \frac{\pi j}{2(N+1)}\right]^2 \qquad j = 1, 2, ---, N \tag{V-5}$$

where the eigenvalues now decrease in magnitude with increasing j. When N is large, λ_j changes slowly with j. Since the values for large j (corresponding to the longest relaxation times from constraint release) contribute very little to G(t), we can replace the summation over j by integration to obtain

$$\frac{G^*(\omega)}{G_N^0} = \frac{8}{\pi^2}\sum_{\substack{odd\\k}}\frac{1}{k^2}I_k(\omega) \tag{V-6}$$

in which, after a change of variables,

$$I_k(\omega) = \frac{2}{\pi}\int_0^{\pi/2}\frac{i\omega\tau_d\, d\xi}{k^2 + i\omega\tau_d + \frac{2}{\Lambda}\cos^2\xi} \tag{V-7}$$

Evaluation of $I_k(\omega)$ gives the result

$$\frac{G^*(\omega)}{G_N^0} = \frac{8}{\pi^2}\frac{\Lambda}{2} i\omega\tau_d \sum_{\substack{odd\\k}}\frac{1}{k^2}\frac{1}{\alpha_k^{1/2}(1+\alpha_k)^{1/2}} \tag{V-8}$$

where

$$\alpha_k = \frac{\Lambda}{2}(k^2 + i\omega\tau_d) \tag{V-9}$$

Values of storage modulus $G'(\omega)$ and loss modulus $G''(\omega)$ can then be obtained by separating Eq. V-8 into its real and imaginary parts (Eq. V-2). Viscosity and recoverable compliance in the large N limit can be obtained from[12)]

$$\eta_0 = \lim_{\omega \to 0} \frac{G'(\omega)}{\omega^2} \tag{V-10 a}$$

$$J_e^0 = \frac{1}{\eta_0^2} \lim_{\omega \to 0} \frac{G'(\omega)}{\omega^2} \tag{V-10 b}$$

with the results:

$$\frac{\eta_0}{G_N^0 \tau_d} = \frac{8}{\pi^2} \left(\frac{\Lambda}{2}\right)^{1/2} \sum_{\substack{\text{odd}\\k}} \frac{1}{k^3} \frac{1}{(1 + \frac{\Lambda}{2} k^2)^{1/2}} \tag{V-11}$$

$$J_e^0 G_N^0 = \frac{1}{2} \left(\frac{2}{\Lambda}\right)^{1/2} \frac{\displaystyle\sum_{\substack{\text{odd}\\k}} \frac{1}{k^5} \frac{1 + \Lambda k^2}{(1 + \frac{\Lambda}{2} k^2)^{3/2}}}{\left[\displaystyle\sum_{\substack{\text{odd}\\k}} \frac{1}{k^3} \frac{1}{(1 + \frac{\Lambda}{2} k^2)^{1/2}}\right]^2} \tag{V-12}$$

These equations reduce to the Doi-Edwards results for $\Lambda \to \infty$ and, for $\Lambda \to 0$ (large constraint release contributions),

$$\frac{\eta_0}{G_N^0 \tau_d} = \frac{8}{\pi^2} \left(\sum_{\substack{\text{odd}\\k}} \frac{1}{k^3}\right) \left(\frac{\Lambda}{2}\right)^{1/2} \tag{V-13}$$

$$J_e^0 G_N^0 = \frac{\displaystyle\sum_{\substack{\text{odd}\\k}} \frac{1}{k^5}}{\left(\displaystyle\sum_{\substack{\text{odd}\\k}} \frac{1}{k^3}\right)^2} \left(\frac{1}{2\Lambda}\right)^{1/2} \tag{V-14}$$

References

1. P. G. de Gennes, J. Chem. Phys. *55*, 572 (1971); see also Chap. VIII of P. G. de Gennes, "Scaling Concepts in Polymer Physics", Cornell University Press, Ithaca, 1979.
2. M. Doi and S. F. Edwards, J. Chem. Soc. Faraday Trans. 2 *74*, 1789 (1978)
3. M. Doi and S. F. Edwards, J. Chem. Soc. Faraday Trans. 2 *74*, 1802 (1978)
4. M. Doi and S. F. Edwards, J. Chem. Soc. Faraday Trans. 2 *74*, 1818 (1978)
5. M. Doi and S. F. Edwards, J. Chem. Soc. Faraday Trans. 2 *75*, 38 (1979)
6. M. Doi, J. Polymer Sci.: Polymer Phys. Ed. *18*, 2055 (1980)
7. M. Doi, J. Polymer Sci.: Polymer Phys. Ed. *18*, 1891 (1980)
8. M. Doi, J. Polymer Sci.: Polymer Phys. Ed. *18*, 1005 (1980)
9. M. Doi and N. Y. Kuzuu, J. Polymer Sci.: Letters *18*, 775 (1980)
10. M. Doi, J. Polymer Sci.: Letters *19*, 265 (1981)
11. W. W. Graessley, J. Polymer Sci.: Polymer Phys. Ed. *18*, 27 (1980)
12. J. D. Ferry, "Viscoelastic Properties of Polymers", 3rd Ed., John Wiley and Sons, New York, 1980
13. W. W. Graessley, Adv. Polymer Sci. *16*, 1 (1974)
14. H.-C. Kan, J. D. Ferry and L. J. Fetters, Macromolecules *13*, 1571 (1980)
15. C. R. Taylor, H.-C. Kan, G. W. Nelb and J. D. Ferry, J. Rheology *25*, 507 (1981)
16. S. Granick, S. Pederson, G. W. Nelb, J. D. Ferry and C. W. Macosko, ACS Polymer Preprints *22*(2), 186 (1981)
17. P. G. de Gennes, J. Phys. *36*, 1199 (1975)
18. W. W. Graessley, T. Masuda, J. E. L. Roovers and N. Hadjichristidis, Macromolecules *9*, 127 (1976)
19. W. W. Graessley, Accts. Chem. Research *10*, 332 (1977)
20. W. E. Rochefort, G. G. Smith, H. Rachapudy, V. R. Raju and W. W. Graessley, J. Polymer Sci.: Polymer Phys. Ed. *17*, 1197 (1979)
21. P. G. de Gennes, Macromolecules *9*, 587 (1976)
22. Ref. 13, Eq. 6.33
23. M. Fukuda, K. Osaki and M. Kurata, J. Polymer Sci.: Polymer Phys. Ed. *13*, 1563 (1975), and earlier articles
24. K. Osaki and M. Kurata, Macromolecules *13*, 671 (1980)
25. W. W. Graessley, Macromolecules *8*, 186 (1975)
26. M. Gottlieb, C. W. Macosko, G. S. Benjamin, K. O. Meyers and E. W. Merrill, Macromolecules *14*, 1039 (1981) and references
27. J. E. Mark, Rubber Chem. Techn. *48*, 495 (1975)
28. J. D. Ferry and H.-C. Kan, Rubber Chem. Tech. *51*, 731 (1978)
29. G. Marrucci and G. de Cindio, Rheologica Acta *19*, 68 (1980); G. Marrucci and J. J. Hermans, Macromolecules *13*, 380 (1980); G. Marrucci, Macromolecules *14*, 434 (1981)
30. C. F. Curtiss and R. B. Bird, J. Chem. Phys. *74*, 2016 (1981); ibid., *74*, 2026 (1981)
31. K. E. Evans, Doctoral Dissertation, The Cavendish Laboratory, University of Cambridge (1980); K. E. Evans and S. F. Edwards, J. Chem. Soc. Faraday Trans. 2, *77*, 1891 (1981); ibid. *77*, 1913 (1981); ibid. *77* 1929 (1981)
31a. R. B. Bird, O. Hassager, R. C. Armstrong and C. F. Curtiss, "Dynamics of Polymeric Liquids" Vol. 2. "Kinetic Theory", John Wiley and Sons, New York, 1977
31b. P. J. Flory, "Principles of Polymer Chemistry", Cornell University Press, Ithaca, 1953
32. R. E. Cohen and N. W. Tschoegl, Intern. J. Polymeric Mater. *3*, 3 (1974)
33. K. E. Evans, J. Chem. Soc. Faraday Trans. 2, to be published
34. W. W. Graessley and J. Roovers, Macromolecules *12*, 959 (1979)
35. G. Marin, E. Menezes, V. R. Raju and W. W. Graessley, Rheologica Acta *19*, 462 (1980)
36. V. R. Raju, E. V. Menezes, G. Marin, W. W. Graessley and L. J. Fetters, Macromolecules *14*, 1668 (1981)
37. R. A. Orwoll and W. Stockmayer, Adv. Chem. Phys. *15*, 305 (1969)
38. J. Klein, Macromolecules *11*, 852 (1978); M. Daoud and P. G. de Gennes, J. Polymer Sci.: Polymer Phys. Ed. *17*, 1971 (1979)
39. J. Klein, Phil. Mag. A *43*, 771 (1981)

40. L. Léger, H. Hervet and F. Rondelez, Macromolecules *14*, 1732 (1981)
41. J. Klein, ACS Polymer Preprints (March, 1981) *22*, 105 (1981)
42. J. D. Ferry, private communication
43. W. W. Graessley and S. F. Edwards, Polymer *22*, 1329 (1981)

Note Added in Proof

A factor 1/2 was omitted from Eq. 88 and in the preceding sentence. The equation should read

$$\Lambda(z) = \left(\frac{\pi^2}{12}\right)^z \frac{1}{z} \qquad (88)$$

This only changes the relationship between $\Lambda(z)$ and z, such as listed in the table following Eq. 101. It does not change the relationship between $\Lambda(z)$ and η_0 and J_e^0 which are also listed in that table.

Solid-State Extrusion of Semicrystalline Copolymers

Michael Dröscher

Institut für Makromolekulare Chemie, Universität Freiburg, Stefan-Meier-Straße 31, D-7800 Freiburg, Federal Republic of Germany

The extrusion of polymers in the solid state is a rather new but important cold forming process. This type of extrusion can be performed either under ram conditions or under hydrostatic pressure. The main experimental parameters which determine the product properties are the temperature of extrusion and the geometry of the die. In general, copolymers have a higher ductility and are more easily extruded than the corresponding homopolymers. The copolymers need lower extrusion pressures and can be extruded to higher extrusion ratios than the homopolymers. The degree of stiffness enhancement upon extrusion in the solid state is lower and the amount of die swell is larger for copolymers. Two types of copolymers have been subjected to investigations: i) copolyethylenes, including short branched polyethylene with different amount and type of side groups as well as an alternating copolymer, ii) segmented block copoly(ether ester)s which are thermoplastic elastomers.

Advances in Polymer Science 47

1 Introduction

Cold forming of polymers has become an important way of shaping solid polymers at temperatures below their softening point which is the glass transition temperature for amorphous materials or the melting temperature for semicrystalline polymers. Typical cold forming procedures are cold drawing, bending, rolling, deep drawing, upsetting, stamping, coining, hydroforming, cold molding, and, with increasing importance, ram or hydrostatic extrusion methods[1].

It should be noted that the first reported solid-state extrusion experiment with polymeric material was performed by metallurgists. Pugh and Low investigated the extrusion behavior of semicrystalline polymers in their treatment of the hydrostatic extrusion of "difficult" metals[2]. They extruded polyethyelene (PE) and polytetrafluoroethylene (PTFE) at extrusion ratios up to 4. In the same year, first investigations on the solid-state extrusion of an amorphous polymer, polystyrene, were reported[3]. Shortly later, Buckley and Long performed room temperature extrusions of several semicrystalline polymers, including low density polyethylene (LDPE), high density polyethylene (HDPE), isotactic polypropylene (PP), and polyoxymethylene (POM)[4]. All of these polymers were found to show a "satisfactory" extrusion behavior in the range of the applied extrusion ratios which were smaller than 6. However, only a weak increase of the axial tensile modulus could be achieved, although the obtained orientation of the samples was similar to that of drawn fibers of the same degree of elongation. For example, the axial tensile modulus of HDPE was increased from 440 MPa, the value of the isotropic sample, to 470 MPa. This value is still very low compared to the theoretical longitudinal tensile modulus of a fully extended PE chain which has been determined to be of the order 240 to 290 GPa[5–7]. Progress on the way to polymers with moduli which come near to the theoretical values was made, when the extrusion temperature had been raised to about 100 °C and the elongation ratio had been increased to values around 30. Now, moduli of up to 70 GPa were observed for HDPE[8].

In the meanwhile, the preparation and characterization of ultrahigh-modulus PE, PP, and POM has become a large field of interest. Contributions were made mainly by Porter[8–10], Ward[11,12], Imada[13], Nakayama[14], and their coworkers. Bigg[15], Ward[16], and Porter[17,18] have published extended reviews on the processes of the ultraorientation of polymers including solid-state extrusion methods. However, these reviews do not discuss the influence of the chemical structure on the extrusion process in detail which comes into focus if copolymers are extruded in the solid state. To fill in this gap, it is the purpose of this article to cover the literature which deals with the solid-state extrusion of copolymers. As this field is very new, it is not possible to discuss the full range of applications for copolymers oriented in the solid state by extrusion, yet. Here, we point out the difference in the extrusion behavior of homopolymers and copolymers and discuss the morphology and the properties of solid state extruded copolymeric materials.

2 Extrusion-Techniques

For the solid-state extrusion several techniques have been developed. They are either based on pure ram extrusion or on hydrostatic methods.

2.1 Ram Extrusion

In ram extrusion experiments, the pressure is applied directly to the semicrystalline billet by use of a piston. Ram extrusion was used by Imada and coworkers for the extrusion of PP[13] and of HDPE[19] and for the coextrusion of HDPE with n-paraffins[20]. Figure 1 gives a schematical representation of the apparatus used for ram extrusion. To decrease the friction of the billet during the extrusion experiment, a lubricant was added. A similar technique, however without lubricant, was applied for the solid-state extrusion of thermoplastic elastomers of the block copoly(ether ester) type by Dröscher and coworkers[21–23].

2.2 Hydrostatic Extrusion

Most of the experiments described in the literature, including the pioneering work by Pugh, were performed as hydrostatic extrusions. For materials like HDPE, PP, and POM this method is considered to be the only one which allows continuous preparation of extrudates, due to the reduced friction between billet and barrel and the decrease in strain hardening[24]. The apparatus also used for hydrostatic extrusion is schematically depicted in Fig. 1.

Recently, this method was improved by adding a haul-off assistance[12] and by the introduction of the so-called split billet technique[25], allowing draw ratios to be applied up to 36 for HDPE.

2.3 Extrusion Parameters

For all solid-state extrusion experiments the temperature of extrusion and the die geometry are the most important parameters. The appropriate temperatures are determined by the nature of the extrudate. The dies used are in most cases of circular shape with a conical inlet. The cone semiangles are most often chosen in the range from 15 to 60°.

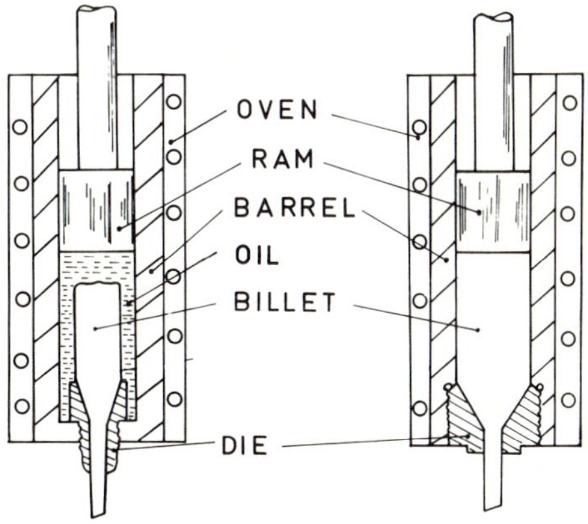

Fig. 1 a, b. Schematic representation of the apparatus used for solid-state extrusion. **(a)** Hydrostatic extrusion, **(b)** ram extrusion

The nominal extrusion ratio or nominal draw ratio (R_N) is determined by the ratio of the billet cross-sectional area to that of the die bore. As it will be discussed below, die swell is often observed in solid-state extrusion experiments[23, 26–28]. Therefore, the actual draw ratio (R_A) is often smaller than the nominal value.

Furthermore, the rate of deformation, often reported as extrusion velocity, and the extrusion pressure are important parameters in extrusion experiments.

3 Criterion for the Solid-State Extrudability of Polymers

Aharoni and Sibilia have established what they call a criterion for the "true" solid-state extrusion[26, 29]. They investigated 13 polymers in their own laboratory and discussed the extrusion behavior of 9 other polymers from the literature. According to their conclusions, a true solid-state extrusion at temperatures significantly below the melting temper-

Table 1. Extrudability of some homopolymers and copolymers

Polymer	abbrev.	T_{ex} °C	T_{α_c} °C	ssex	T_m °C	Ref.
high density polyethylene	HDPE	80–145	~88	yes	139–148	17, 26, 30, 34
short branched polyethylene (0.5 mol-% methyl side groups)	PEMC	90–110	~88	yes	127–129	27, 28, 30, 34, 35
short branched polyethylene (0.1 mol-% butyl side groups)	PEBC	90	~88	yes	?	28, 30, 34
low density polyethylene	LDPE	25	~70	yes	106	4, 45
isotactic polypropylene	PP	70–150	≤100	yes	176	26, 37, 38, 39
polyoxymethylene	POM	174–182	102	yes	182	26, 34
poly(vinyl fluoride)	PVF	165–196	110	yes	196	26, 40
poly(vinylidene fluoride)	PVDF	162–175	90	yes	165, 175	26, 40, 41
polycaprolactam	Nylon 6	80–180	–[a]	yes[b]	233	26, 37, 42, 43
poly(hexamethylene adipamide)	Nylon 66	200	–	yes[b]	272	26, 37, 42
poly(ethylene terephthalate)	PET	none	–	no	258	26
poly(butylene terephthalate)	PBT	none	–[c]	no	223	26, 37, 44
poly(butylene terephthalate-co-poly(oxytetra-methylene)terephthalate)	PBTPOTM	45–190	–	yes[b]	162–216	21, 22, 23
poly(ethylene-co-chlorotri-fluoroethylene) (1:1)	PECTFE	170–228	130	yes	236	26, 33

T_{ex} extrusion temperature or oven temperature
T_{α_c} temperature of the crystalline α-relaxation
ssex solid-state extrudability
T_m melting temperature of the extrudate or, if not extrudable, of melt-crystallized material
[a] crystalline phase transition
[b] extruded to rather low draw ratios (under 10)
[c] at room temperature a stress-induced phase transition between two crystalline modifications exists

ature of the crystalline phase of the sample yields smooth, clear or translucent extrudates of draw ratios larger than 14, with apparently no die swell. They found that, if this was the case, the success of the extrusion experiment was not effected by the entrance angle of the die, as long it was in the range of 15 to 45°. Then, the same results with respect to the nature and appearance of the extrudates were obtained[26]. It was concluded that the requirement for a successful solid-state extrusion is the existence of a crystalline relaxation process (α_c) below the temperature at which the extrusion is performed. They pointed out that an abrupt increase of the thermal expansion coefficient takes place at the temperature T_{α_c} which leads to some breakdown of the intracrystalline forces within the crystalline fraction of the polymer manifesting itself by a reduction in the modulus.

This effect has been observed in several polymeric materials, e.g. PE[30], PP[30], PTFE[31], POM[32], and poly(ethylene-co-chlorotrifluoroethylene) (PECTFE)[33]. All of these materials follow the extrudability criterion.

In Table 1 data on the extrusion behavior of some polymers and copolymers are summarized. Many of these data refer to the paper by Aharoni and Sibilia[26]. Some of the polymers listed in Table 1, namely the nylons and the copoly(ether ester)s, do not fulfill the requirements defined by Aharoni and Sibilia. Especially, the maximum achievable draw ratio seems to be less than ten for these systems. However, together with other workers in this area, we think that the term "solid-state extrusion" covers a wider field of interest than indicated by Aharoni and Sibilia.

4 Solid-State Extrusion of Copolyethylenes

In this chapter the solid state extrusion of different grades of polyethylene is discussed. The term copolyethylene stands as well for short and long branched PE as for the nearly alternating 1:1 copolymer poly(ethylene-co-chlorotrifluoroethylene) (PECTFE). It is well known that even HDPE contains a certain amount of short branches. Therefore, it is of interest to note that already one butyl side group per thousand main chain carbon atoms effects the solid state extrusion properties of PE remarkably.

Not all polymers which are subject to this chapter have been characterized adequately in the original papers. Especially the commercial materials might contain additives which could be of more importance for the extrusion behavior than the side groups.

4.1 Low Density Polyethylene

Low Density Polyethylene (LDPE) was extruded by Buckley and Long at room temperature[4]. They reported "satisfactory" extrusion for nominal draw ratios of 3, 5, and 7. The observed diametral die swell amounted to 20%, 12%, and 11%, respectively. For the same set of experiments the extrusion pressure increased from 31.5 MPa to 94.5 MPa. These values are about 30% lower than for the extrusion of HDPE under similar conditions.

The orientation of the extruded LDPE samples was similar to the orientation of cold drawn material. The tensile modulus increased only slightly from 107 MPa for the isotropic sample to 174 MPa for the sample with the highest draw ratio.

4.2 Short Branched Polyethylene

4.2.1 Description of the Material and Extrusion Experiments

Hope and coworkers have investigated the solid-state extrusion under hydrostatic pressure of different grades of linear PE[27, 28, 35]. Table 2 lists the amount and type of side groups, the average molecular weights, and the melt flow index (MFI) of the extruded materials. All samples are of the "Rigidex" type (B.P. Chemicals Ltd.) and were chosen by the authors to be directly comparable with each other in terms of the MFI. But for the butyl copolymer the MFI lies in all cases within the optimum range from about 5 to 10[46].

These extrusion experiments were carried out with circular dies at temperatures between 90 and 120 °C. The cone semi-angle of the die was 15° and the exit-bore diameter varied between 2.5 and 23.8 mm. The extrusion was supported by a haul-off stress of less than 10 MPa.

As shown in Fig. 2 for two homopolymer samples and one copolymer, the polymer grade had a strong influence on the extrusion behavior. At an extrusion temperature of 110 °C the highest extrusion pressure was needed for the high molecular weight homopolymer sample. The difference in extrusion pressure increased with increasing draw ratio. The copolymer was more easily extruded than the homopolymers. Consequently, the authors observed a significant variation of the maximum extrusion ratio

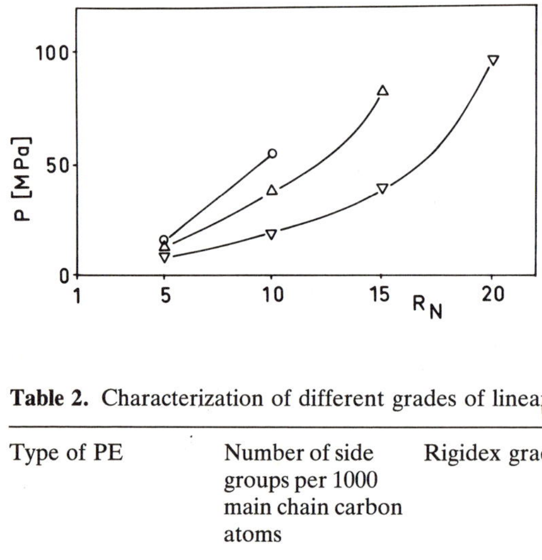

Fig. 2. Comparison of the solid-state extrusion behavior of different grades of linear polyethylene. Extrusion temperature 110 °C, extrusion rate 1 mm/min. (○) H 020-54, (△) R 006-60, (▽) R 40 (for sample description see Table 2)[35]

Table 2. Characterization of different grades of linear polyethylene[27, 28, 35]

Type of PE	Number of side groups per 1000 main chain carbon atoms	Rigidex grade	$\overline{M}_w$	$\overline{M}_n$	MFI[b]
homopolymer	[a]	R 50	101 450	6 180	5.5
homopolymer	[a]	R 075-60	68 200	14 150	8.0
homopolymer	[a]	R 006-60	135 000	25 500	11.8
homopolymer	[a]	H 020-54 P	312 000	33 000	9.6
methyl copolymer	5	R 40	93 600	9 600	4.0
methyl copolymer	5	R 85	80 800	10 700	9.0
butyl copolymer	1	R 002-55	167 500	19 400	0.2

[a] not reported
[b] melt flow index

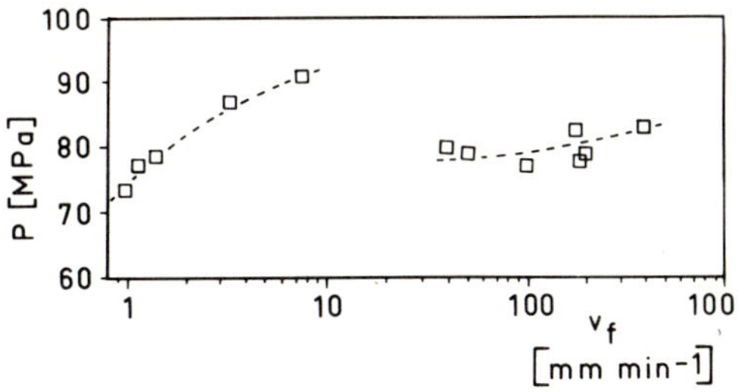

Fig. 3. Variation of extrusion pressure with extrudate velocity. Extrusion temperature 90 °C, nominal draw ratio 15, die diameter 15.5 mm, sample R 40 (for sample description see Table 2). Régimes at lower and higher velocities are isothermal and adiabatic, respectively[27)]

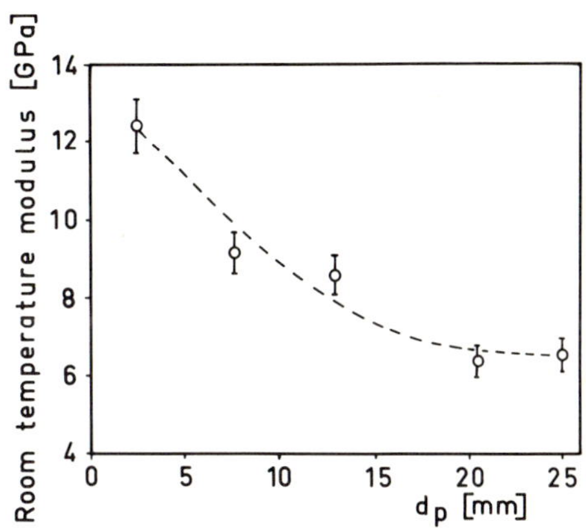

Fig. 4. Variation of the axial tensile modulus with product diameter. Extrusion temperature 100 °C, nominal draw ratio 10, sample R 40 (for sample description see Table 2)[27)]

obtainable from each PF grade: The highest extrusion ratio at which unflawed products were obtained under any set of extrusion conditions decreased from 25 for the methyl copolymer R 40 to 20 and 15 for the low molecular weight PE R 006-60 and the high molecular weight PE H 020-54 P, respectively.

Hope and Parsons employed the methyl copolymer R 40 for an investigation of the solid-state extrusion with large die diameter[27)]. As shown in Fig. 3, the extrusion of a rod of 15.5 mm diameter could be carried out at different extrusion velocities. At low velocities the pressure increased monotonically with velocity. At higher velocities an instability occurred and the extrudate appeared to be partially melted. However, if the pressure control had been readjusted to a lower value at the onset of instability, the steady-state extrusion would have worked again in a second régime. The reason for the existence of two distinct extrusion régimes was attributed to a transition from a completely isothermal to a predominantly adiabatic régime. Here the heat transfer was too slow to transport the heat of deformation out of the sample into the tooling, resulting in a rise in the actual extrusion temperature.

4.2.2 Properties of the Extrudates

Figure 4 demonstrates the influence of the product diameter on the product properties. For a given nominal draw ratio of 10 the axial tensile modulus, as determined from a three-point bending test, decreased with increasing product diameter. The authors attrib-

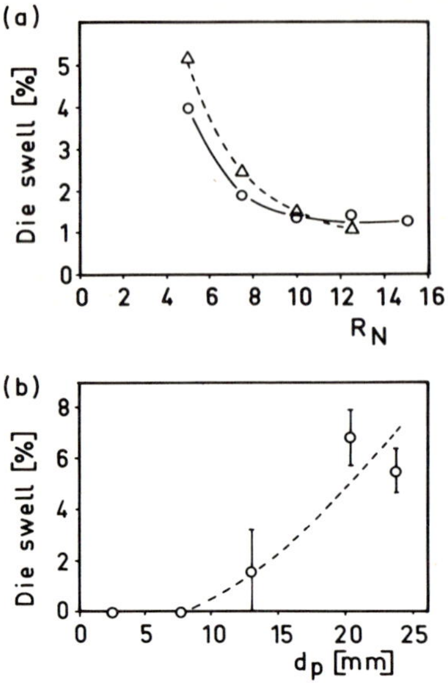

Fig. 5 a, b. Percentage die swell of products. **(a)** Effect of draw ratio, **(b)** effect of product diameter. Sample R 40 (for sample description see Table 2). Extrusion velocity (○) 0.1 m/min, (△) 3 m/min[27)]

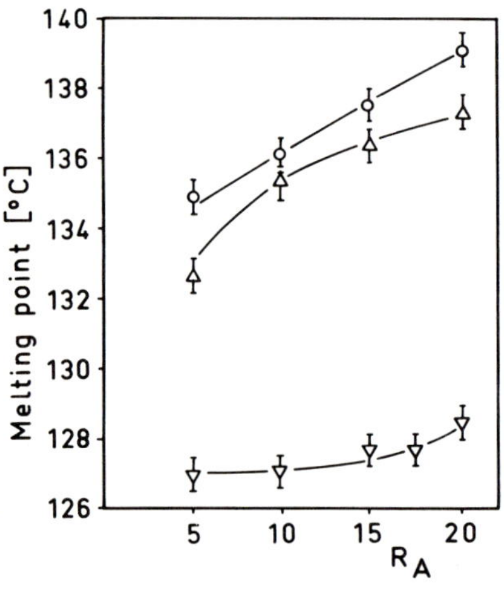

Fig. 6. Differential scanning calorimetry melting points of extrudates (at a heating rate of 10 K/min). Extrusion temperature 100 °C, die diameter 2.5 mm. (○) H 020-54, (△) R 006-60, (▽) R 40 (for sample description see Table 2)[35)]

uted this effect to the increase in the extrudate temperature due to decreasing heat transfer for thicker products[27)].

The degree of diametrical die swell depended strongly on the draw ratio and on the product diameter, as shown in Fig. 5. The influence of the extrusion velocity on the die swell was rather small. For extrusions at draw ratios larger than 10 and for extrudates with diameters smaller than 8 mm the authors did not observe any die swell.

In the described set of experiments Hope and Parsons observed three types of product flaws, internal cracks, discontinuous cracks, and continuous surface cracks. Internal cracks were found to be due to voids in the billet material and could be avoided by the use of void-free billets. Discontinuous cracks were attributed to environmental stress cracking due to the extrusion fluid. This crack process could be reduced in effect by increasing the extrusion velocity. The continuous surface cracks were more difficult to avoid. They were tentatively attributed to the relaxation of the transverse elastic component of the deformation, which was found to be also responsible for the die swell. At low extrusion ratios the ductility of the products was sufficiently large to relieve the stresses by die swell. For the materials which had been subjected to higher extrusion ratios these stresses were relieved by cracking, because of the weakness of the highly ordered structure in the transverse direction.

The thermal properties of the extrudates depended on the extrusion conditions and on the chemical structure of the sample. This is shown in Fig. 6 for the dependence of the melting temperature on the draw ratio. The small difference in the melting behavior of the two homopolymer samples was attributed to the higher orientation of the noncrystalline regions in the high molecular weight sample compared to the low molecular weight material[35]. Because of the well known melting point depression of copolymers, the copolymer sample R 40 melted at a lower temperature than the homopolymers. But, in contrast to the homopolymer samples, the melting temperature did not increase upon extrusion.

As shown in Fig. 7, the axial room temperature tensile modulus depended both on the chemical structure of the extrudate and the draw ratio applied. In the medium-low molecular weight range it was found for the homopolymers that, to a good approximation, the relationship between the modulus and the draw ratio was independent of molecular weight. For the high molecular weight homopolymer and for the copolymer R 40 significantly lower moduli were observed over the entire range of R_A[35]. This result was explained by the so-called crystalline bridge model which had been derived for highly oriented materials by Gibson et al.[16, 47]. In this model the crystalline phase is the continuous phase and the crystalline regions are connected by crystalline links. In terms of this model, the decrease in the modulus of the copolymer sample was due to a lower crys-

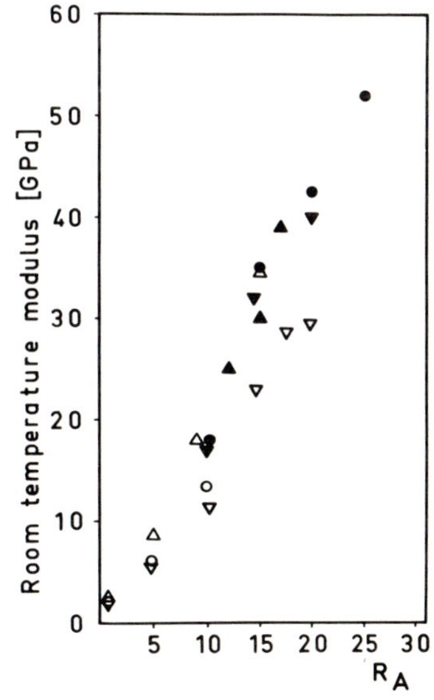

Fig. 7. Comparison of axial room temperature tensile moduli for linear polyethylene extrudates. Extrusion temperature 100 °C, die diameter 2.5 mm. (▽) R 40, (○) H 020-54 P, (△) R 006-60, (▼) R 25, (●) R 50, (▲) R 140-60 (for sample description see Table 2, for R 25 and R 140-60 see Ref. 35)[35]

Table 3. Tensile properties and die swell for different solid-state extruded linear polyethylenes[28]

Type of PE	Rigidex grade	Room temperature tensile modulus GPa $R_N = 9$	$R_N = 15$	Die swell % $R_N = 9$	$R_N = 15$
homopolymer	R 50	12.7	23.4	0.1	0.6
homopolymer	R 075-60	15.5	23.8	0.0	1.9
methyl copolymer	R 40	6.03	11.8	1.9	1.3
methyl copolymer	R 85	6.77	10.6	1.5	2.6
butyl copolymer	R 002-55	5.11	–	3.9	–

R_N nominal draw ratio

tallinity and lower orientation as compared to the homopolymer sample of medium molecular weight.

The influence of the butyl side groups on the tensile properties of the extrudates was even stronger than that of the methyl side groups. This is demonstrated in Table 3[28] for extrudates where the nominal draw ratios were 9 and 15. The strong decrease of the axial room temperature modulus with increasing size of the side groups went parallel with the increase of the amount of die swell.

Thus, Hope and his coworkers concluded for the solid-state extrusion of linear polyethylenes that increasing the molecular weight reduced both the maximum degree of deformation and the tensile properties of the product. The solid-state extrusion of copolymers, on the other hand, increased the maximum degree of deformation. The degree of stiffness enhancement upon extrusion and the melting points of the products were reduced.

4.3 Poly(ethylene-co-chlorotrifluoroethylene)

Aharoni and Sibilia investigated the solid-state extrusion behavior of PECTFE, using the material sold as "Halar" by Allied Chem. Corp.[26]. This copolymer is a highly alternating 1:1 system. Modena et al. found by IR methods that the amount of alternating sequences is 93 to 97%, depending on the temperature at which the synthesis has been carried out[48]. For PECTFE an α-transition at 130 °C was found[33] which was attributed to the crystalline regions, because its amplitude increased with rising crystallinity of the sample. The authors applied extrusion temperatures between 170 and 228 °C, using dies with an exit-bore diameter of 2.54 mm and entrance angles in the range from 15 to 45°. The velocity of extrusion was about 38 mm/min. Unfortunately, the only physical property reported for the extrudates was the thermal behavior.

Poly(ethylene-co-tetrafluoroethylene) does not reveal a crystalline transition and could not be extruded in the solid state[26].

5 Solid-State Extrusion of Segmented Poly(ether ester)s

Thermoplastic elastomers combine certain desirable properties of elastomers – elastic recovery and energy absorption – with the processing convenience of thermoplastics. These properties are based on the phase-separation behavior of the materials which consist of so-called "soft" segments with a low glass transition temperature and of "hard" segments with a high glass transition temperature or melting temperature. The hard component provides the reversible network structure and ensures the dimensional stability of the product by minimizing the cold flow of the sample.

The segmented block copoly(ether ester)s, based on poly(butylene terephthalate) (PBT) and on polyoxytetramethylene (POTM) are typical examples for this class of thermoplastic elastomers[49–51]. They can be described as random copolyesters of terephthalic acid with 1,4-butanediol and α-hydro-ω-hydroxypolyoxytetramethylene. A certain fraction of the ester sequences crystallizes and forms the hard phase whereas the remaining ester segments mix with the ether segments and build up the soft matrix.

The first extrusion experiments with this type of material was carried out under ram conditions by Dröscher and Regel[21]. Applying a rectangular die, they obtained a well oriented sample at 190 °C, which was about 22 °C below the melting temperature of the isotropic poly(ether ester) sample. Much more extended work was subsequently carried out with circular dies where the conical inlet had a semiangle of 41°[22, 23].

5.1 Description of the Materials and Extrusion Experiments

The materials investigated by Dröscher and coworkers had been synthesized by the Toyobo Co. Katata Research Center in Otsu City, Japan, as described by Hoeschele[51]. The samples varied in the average degree of polymerization of the "hard" PBT segments, as given in Table 4. The POTM block length had been kept constant at about 14 units. At this length, the POTM segments did not tend to crystallize, yet[52]. The intrinsic viscosities were for all three materials of the order of 1.5 dl/g in phenol/tetrachloroethane mixtures (60/40 p/w) at 30 °C.

To prepare billets for the solid-state extrusion experiments, the materials were injection molded at 15 to 30 °C above the respective melting temperature. The billet materials

Table 4. Characterization of block copoly(ether ester) samples (injection-molded)[23]

Material	x_H	w_H	$\overline{P}_{n,H}$	$\overline{P}_{n,S}$	T_{im}	T_m	Δh_m	α_m
A	0.79	0.45	4.8	14	190	162	12	0.19
B	0.84	0.54	6.3	14	210	197	24	0.32
C	0.92	0.72	12.5	14	230	216	39	0.39

x_H mole fraction of the PBT segments, w_H weight fraction of the PBT segments, $\overline{P}_{n,H}$ and $\overline{P}_{n,S}$ average degree of polymerization for the PBT and POTM segments, respectively, T_{im} temperature of the injection molding process, T_m melting temperature, Δh_m heat of melting, α_m weight fraction of the crystallized PBT (α_m = 1: all PBT segments crystallized)

Table 5. Solid-state extrusion experiments with the poly(ether ester) PBTPOTM[23]

Material	Temperature of extrusion °C	Nominal draw ratio	Rate of extrusion mm^3/s	Die swell %
A	48–118	4	0.4	40–34
B	75–170	4	0.4	29–26
C	50–190	4	0.4	22–12
C	180	2–7	0.4	12–13
C	180	4	0.4–383	12–24

Table 6. Extrusion conditions for the poly(ether ester) C. Nominal draw ratio 4 (for sample description see Table 4)[23]

Temperature of extrusion °C	Rate of extrusion mm^3/s	Pressure MPa
138	0.4	20
	1.9	33
	3.8	40
190	0.4	18
	3.8	20

thus obtained are characterized in Table 4. The billets were 6.9 mm in diameter and 90 mm in length to fit into the chamber of the capillary rheometer. They did not show any orientation from the molding procedure.

In Table 5 the experimental conditions for the extrusion experiments are compiled. In all cases die swell occurred. The largest die swell (40%) was observed for material A at an extrusion temperature of 48 °C. In all cases the die swell increased with decreasing temperature. As shown in Table 5 for material C, the degree of die swell was independent of the extrusion ratio. This is in contrast to the results on copolyethylenes (see Chap. 4). The die swell depended on the extrusion rate, however. This led the authors to the conclusion that the amount of reversible deformation of the extruded material in the die was determined by the volume fraction of the soft matrix. This volume fraction of the soft matrix increased with decreasing hard segment lengths and falling extrusion temperature.

Most experiments were run at an extrusion rate of 0.4 mm^3/s. At this rate, little influence of the extrusion temperature on the extrusion pressure was observed as the data in Table 6 indicate for material C.

The extrudates obtained under the described conditions were 30 to 40 cm long, fracture-free and of smooth and shiny surface.

5.2 Morphology of the Extrudates

Figure 8 shows four wide-angle X-ray diagrams with fiber geometry for material C. The extrusion temperature was varied from 180 to 65 °C from (a) to (d). In all four cases the reflections could be indexed according to the crystallographic cell parameters of the so-called α-structure of PBT[53]. The samples differed in the degree of orientation of the crystalline regions. From the intensity distribution of the equatorial reflections the authors determined the second-moment orientation function f_c[20] which becomes equal to 1 for complete orientation and -0.5 for orientation in the perpendicular directions. For the extrudates shown in Fig. 8, f_c was found to decrease from 0.97 to 0.86, if the extrusion temperature was lowered from 180 to 65 °C at a nominal draw ratio of 4. Changing the draw ratio at a constant extrusion temperature also influenced the crystallite orientation. At 180 °C, f_c varied from 0.78 to 0.98, if the draw ratio was increased from 2 to 7.

Small-angle X-ray studies gave rather sharp small-angle scattering peaks in all cases. Second-order peaks were never observed. From the Bragg angle of the peak maxima the authors calculated the long spacings which are listed in Table 7. The long period clearly depended on the extrusion conditions. For all three materials the long period increased with rising extrusion temperature. The largest effect occurred for material C. Changing the rate of extrusion or the nominal draw ratio at an extrusion temperature of 180 °C did not influence the long period. Thus, the long period was only determined by the undercooling, the temperature difference between the melting temperature of the respective material and the extrusion temperature.

The melting temperatures, as determined by differential scanning calorimetry, were nearly independent of the extrusion conditions. Similarly, the melting temperatures of

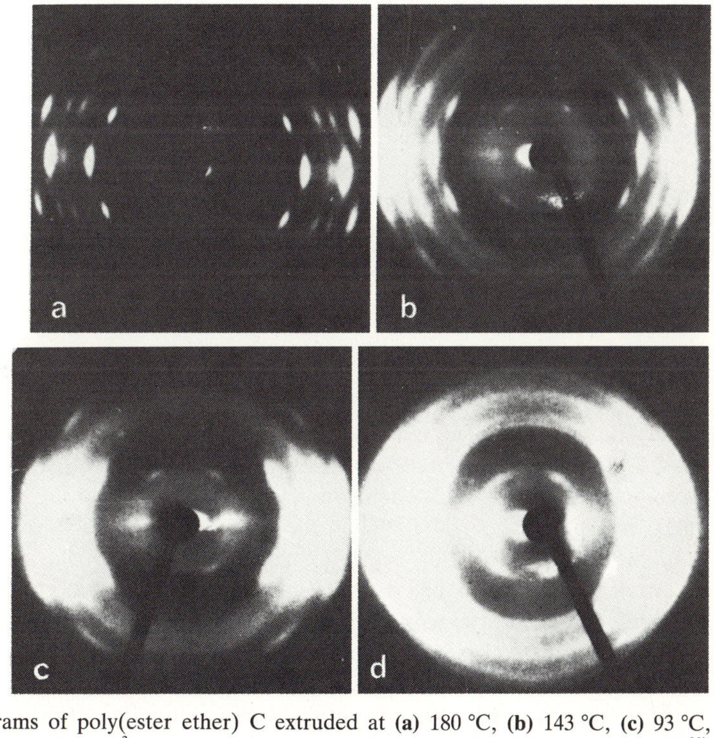

Fig. 8 a–d. Fiber diagrams of poly(ester ether) C extruded at **(a)** 180 °C, **(b)** 143 °C, **(c)** 93 °C, **(d)** 65 °C. Rate of extrusion 0.4 mm^3/s, nominal draw ratio 4 (for sample description see Table 4)[23]

Table 7. Small-angle X-ray data for poly(ether ester) extrudates (for sample description see Table 4)[23]

Material	Temperature of extrusion °C	Rate of extrusion mm^3/s	Nominal draw ratio	Long period nm
A	118	0.4	4	13.3
	103			12.3
	90			11.5
	68			11.0
	48			10.8
B	170	0.4	4	14.9
	150			12.0
	130			12.0
	112			12.0
	94			11.5
	75			11.5
C	190	0.4	4	14.4
	180	0.4–384	2–7	12.7 ± 0.3
	160	0.4	4	11.5
	140			11.2
	120			10.1
	90			9.7
	60			9.1

isotropic samples did not depend much on the annealing conditions[52]. As shown in Fig. 9, the heat of melting decreased with increasing undercooling in all three cases. For the purpose of comparison, the heats of melting of the isotropic billets were indicated by arrows. The higher heats of melting observed at small undercoolings were attributed by the authors to annealing effects. The heats of melting of the extrudates obtained at low temperatures were lower than those of the corresponding billets. Because of this fact, the authors concluded that the overall crystallinity of the extrudates was determined by *two* opposing effects, the annealing process dominating at high temperatures and the destruction of crystallites due to deformations dominating at low temperatures.

5.3 Mechanical Properties

All tensile measurements were performed by the authors with microtomed ribbons of 0.1 mm thickness at ambient temperature. In Fig. 10 typical stress-strain curves for all three poly(ether ester) materials are plotted. All samples were extruded at a common undercooling of 60 °C. The initial tensile modulus increased from 14 MPa for material A to 62 MPa and 208 MPa for B and C, respectively.

As demonstrated in Fig. 11 for material B, most of the tensile deformation was reversible, even at strains of over 40%. In Fig. 11 for each cycle a new sample was employed. In cases where the same specimen had been subjected to repeated load cycling, the authors observed a substantial amount of strain hardening.

The highest moduli were obtained, at extrusion temperatures in the range 100–130 °C. A further increase of the extrusion temperature led to a decrease of the tensile modulus. The authors attribute this effect again to the annealing process at high extrusion temperatures by which the number of taut molecules should decrease.

Figure 12 shows the influence of the nominal draw ratio on the tensile properties for the poly(ether ester) C. The initial tensile modulus was nearly independent of the draw ratio. A higher strains the modulus increased proportionally to the draw ratio. As can be seen from Fig. 13, the effect of the extrusion velocity on the tensile properties was rather small.

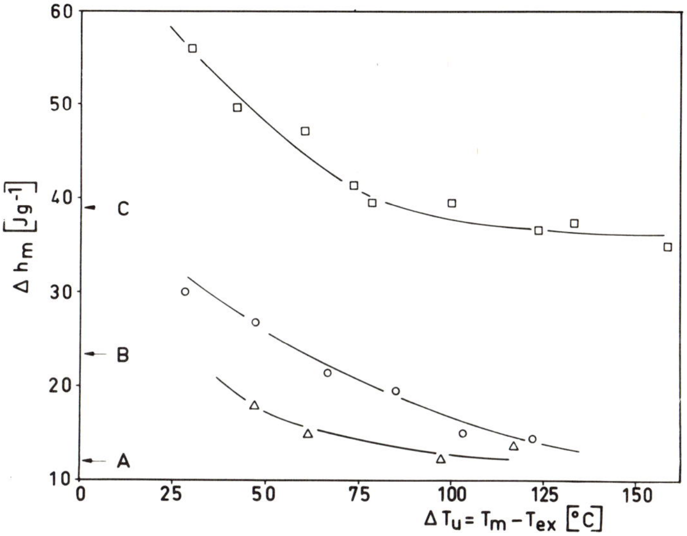

Fig. 9. Heat of melting of extrudates vs. supercooling for poly(ether ester)s. (□) material C, (○) B, (△) A. Arrows denote the heats of melting of the isotropic samples prior to extrusion. Rate of extrusion 0.4 mm^3/s, nominal draw ratio 4 (for sample description see Table 4)[23)]

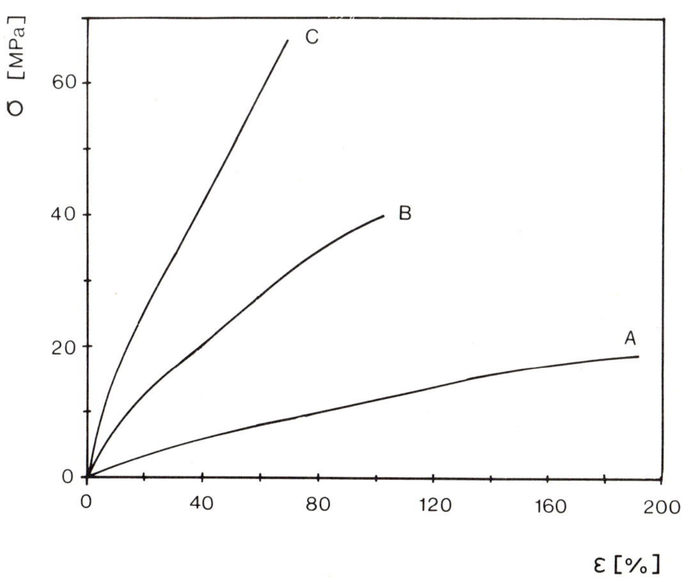

Fig. 10. Stress-strain curves of extrudates for poly(ether ester)s A, B and C. Initial sample length 25 mm, strain rate 1 mm/s, 25 °C, nominal extrusion ratio 4, undercooling 60 °C (for sample description see Table 4)[23)]

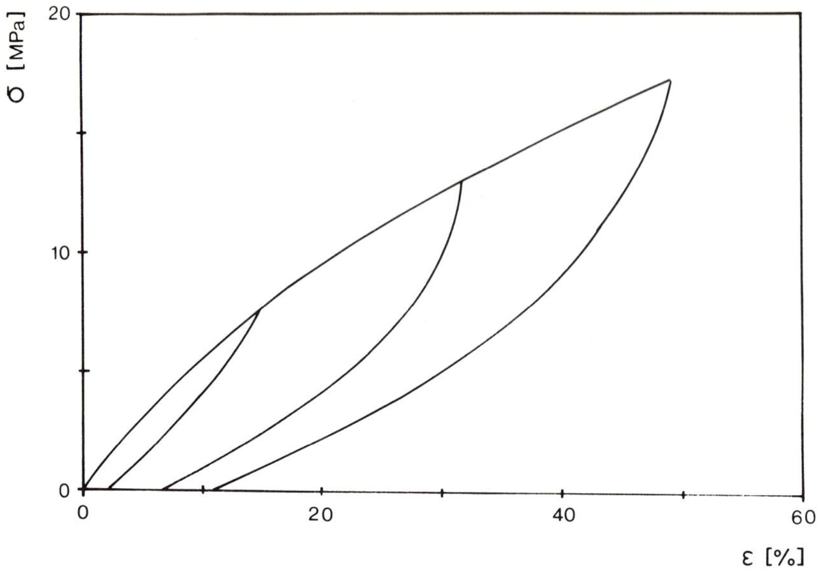

Fig. 11. Stress-strain hysteresis for poly(ether ester) B. Initial sample length 25 mm, strain rate 1 mm/s, 25 °C. For each cycle a new sample was used. Rate of extrusion 0.4 mm^3/s, nominal draw ratio 4, extrusion temperature 110 °C (for sample description see Table 4)[23)]

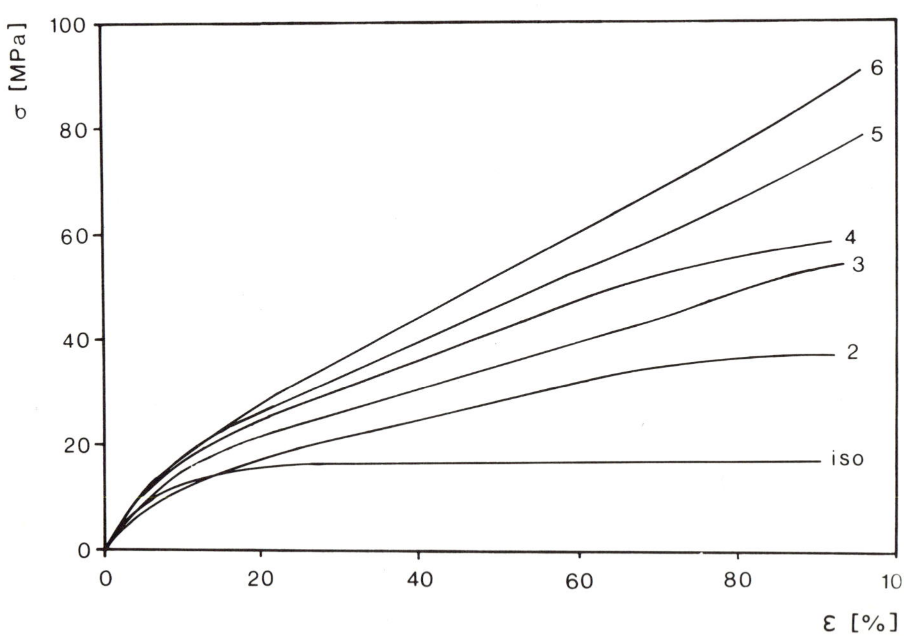

Fig. 12. Stress-strain curves for poly(ether ester) C. Parameter nominal draw ratio. Initial sample length 25 mm, strain rate 1 mm/s, 25 °C. Rate of extrusion 0.4 mm^3/s, extrusion temperature 180 °C (for sample description see Table 4)[23)]

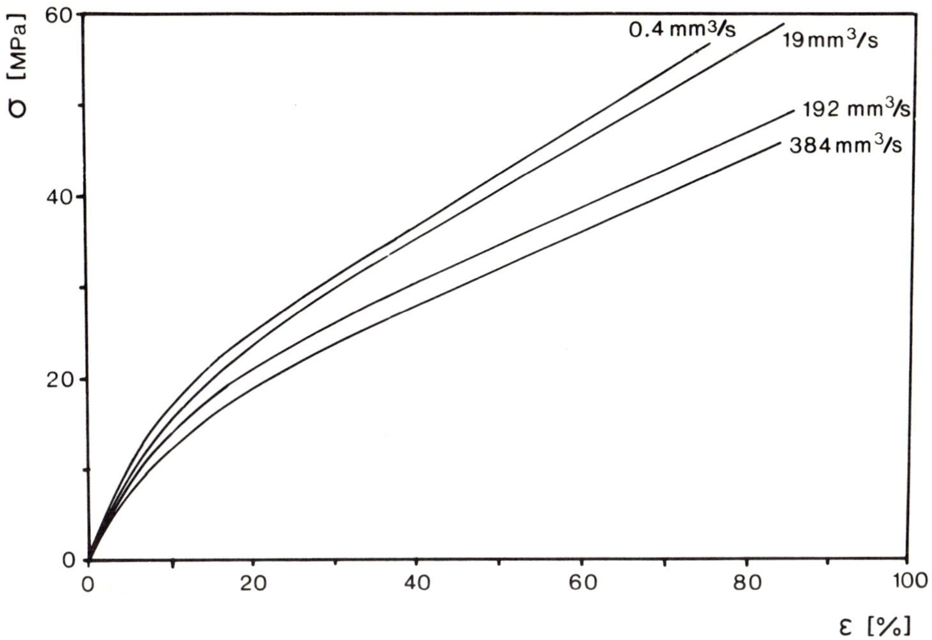

Fig. 13. Stress-strain curves for poly(ether ester) C. Parameter rate of extrusion. Initial sample length 25 mm, strain rate 1 mm/s, 25 °C. Nominal draw ratio 4, extrusion temperature 180 °C (for sample description see Table 4)[23)]

For dynamical-mechanical investigations of the extrudates Dröscher and coworkers used an automatic torsion pendulum. This instrument does not work at constant frequency. The frequency was allowed to range from 0.1 to 10 Hz. Therefore, the curves for the storage modulus G' and the logarithmic decrement Λ plotted in Fig. 14 could deviate slightly from similar curves obtained at 1 Hz. Measurements were only conducted up to 80 °C because of the strong elastic recovery effect which was observed above 100 °C.

Similarly to the behavior of isotropic poly(ether ester)s[52)] the amplitude and position of the relaxation peaks in the loss curve of the extrudates were influenced by the composition of the amorphous phase. This is determined by the concentration and the degree of polymerization of the ester segments. For the extrudates the observed effect was pronounced only in the case of material C. Here, the glass transition temperature, as determined from the maximum of the so-called α-relaxation peak, increased linearly with decreasing extrusion temperature from − 4 °C to 17 °C[55)]. For the materials A and B the glass transition temperatures were found to be − 59 and − 50 °C, respectively, independently of the extrusion conditions.

The peak width of the α-relaxation process is a probe for the heterogeneity of the chain environment in the amorphous phase of the sample. For material C a variation of the peak width was found if the extrusion conditions had been changed. This effect was attributed to the changing number of tie molecules under different extrusion conditions.

The β-relaxation process of PBT at about − 100 °C[56, 57)] was only observed for material C. It did not occur in the two other materials where only a γ-transition was found at about − 150 °C. The γ-relaxation was attributed to local motions in the POTM tetramethylene sequences which is normally called β-transition in the case of pure POTM[58)].

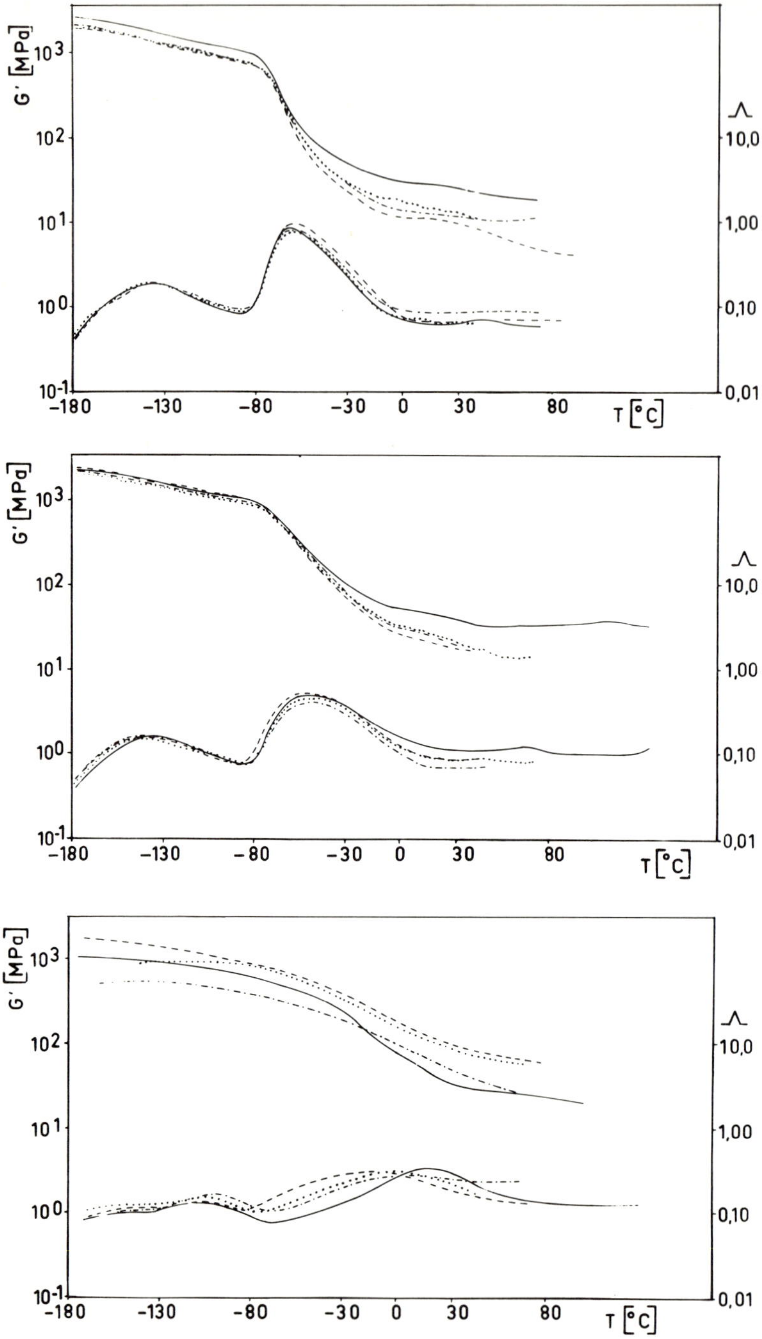

Fig. 14. a–c. Shear modulus and logarithmic decrement vs. temperature for poly(ether ester) extrudates. Parameter extrusion temperature. Solid curves denote materials prior to extrusion. **(a)** material A: (– – –) 118 °C, (· · ·) 68 °C, (– · –) 48 °C, **(b)** material B: (– – –) 170 °C, (· · ·) 130 °C, (– · –) 94 °C, **(c)** material C: (– – –) 190 °C, (· · ·) 116 °C, (– · –) 64 °C. Rate of extrusion 0.4 mm^3/s, nominal draw ratio 4 (for sample description see Table 4)[23]

6 Conclusion

Up to now the number of materials which have been subjected to solid-state extrusion is very small. The deformation mechanisms and the correlation of the deformation conditions with the extrudate properties are far from being understood. This holds especially for multiphase polymers. We have reported a field of polymer science which seems to be still in its beginning.

7 References

1. Shaw, M. T.: Cold forming of polymeric materials, in: Ann. Rev. Material Sci. (eds.) Huggins, R. A., Bube, R. H., Vermilyea, D. A., Vol. 10, p. 19, Palo Alto, Annual Reviews Inc. 1980
2. Pugh, H. Ll. D., Low, A. H.: J. Inst. Metals *93,* 201 (1964/65)
3. Holliday, L. et al.: Nature *202,* 382 (1964)
4. Buckley, A., Long, H. A.: Polym. Eng. Sci. *9,* 115 (1969)
5. Frank, F. C.: Proc. Roy. Soc. (London) *A 319,* 127 (1970)
6. Sakurada, I., Kaji, K.: J. Polym. Sci. Polym. Symp. *31,* 57 (1970)
7. Tashiro, K., Kobayashi, M., Tadokoro, H.: Macromolecules *10,* 413 (1977)
8. Mead, W. T., Porter, R. S.: Intern. J. Polym. Mater. *7,* 29 (1979)
9. Southern, J. H., Porter, R. S.: J. Appl. Polym. Sci. *14,* 2305 (1970)
10. Shimada, T. et al.: J. Appl. Polym. Sci. *26,* 1309 (1981)
11. Gibson, A. G. et al.: J. Mater. Sci. *9,* 1193 (1974)
12. Coates, P. D., Ward, I. M.: Polymer *20,* 1553 (1979)
13. Nakamura, K., Imada, K., Takayanagi, M.: Intern. J. Polym. Mater. *2,* 71 (1972)
14. Nakayama, K., Kanetsuna, H.: Kobunshi Kagaku *30,* 713 (1973)
15. Bigg, D. M.: Polym. Eng. Sci. *16,* 725 (1976)
16. Capaccio, G., Gibson, A. G., Ward, I. M.: Drawing and hydrostatic extrusion of ultra-high modulus polymers, in: Ultra-High Modulus Polymers (eds.) Ciferri, A., Ward, I. M., p. 1, London, Applied Science 1979
17. Zachariades, A. E., Mead, W. T., Porter, R. S.: Recent developments in ultramolecular orientation of polyethylene by soldid-state extrusion, in: Ultra-High Modulus Polymers (eds.) Ciferri, A., Ward, I. M., p. 77, London, Applied Science 1979
18. Watts, M. P. C., Zachariades, A. E., Porter, R. S.: Contep. Top. Polym. Sci. *3,* 297 (1979)
19. Imada, K., Takayanagi, M.: Intern. J. Polym. Mater. *2,* 89 (1973)
20. Maruyama, S., Imada, K., Takayanagi, M.: Intern. J. Polym. Mater. 2, 105 (1973)
21. Dröscher, M., Regel, W.: Polym. Bull. *1,* 551 (1979)
22. Dröscher, M., Bandara, U., Regel, W.: Prepr. IUPAC Symp. on Macromolecules, Vol. III, 141, Florenz 1980
23. Bandara, U., Dröscher, M.: Angew. Makromol. Chem., in press
24. Alexander, J. M.: Mater. Sci. Eng. *10,* 70 (1972)
25. Zachariades, A. E., Griswold, P. D., Porter, R. S.: Polym. Eng. Sci. *19,* 441 (1979)
26. Aharoni, S. M., Sibilia, J. P.: Polym. Eng. Sci. *19,* 450 (1979)
27. Hope, P. S., Parsons, B.: Polym. Eng. Sci. *20,* 589 (1980)
28. Hope, P. S., Parsons, B.: ibid. *20,* 597 (1980)
29. Aharoni, S. M., Sibilia, J. P.: J. Appl. Polym. Sci. *23,* 133 (1979)
30. McGrum, N. G., Read, B. E., Williams, G.: Anelastic and Dielectric Effects in Polymeric Solids, p. 388, London – New York – Sydney, John Wiley & Sons 1967
31. Goldman, A. Ya., Grinman, A. N., Dunichev, Yu. F.: Polym. Mech. *11,* 340 (1975)
32. Coates, P. D., Ward, I. M.: J. Polym. Sci. Polym. Phys. Ed. *16,* 2031 (1978)
33. Sibilia, J. P., Schaffhauser, R. J., Roldan, L. G.: J. Polym. Sci., Polym. Phys. Ed. *14,* 1021 (1976)

34. Enns, J. B., Simha, R.: J. Macromol. Sci. Phys. Ed. *B 13,* 25 (1977)
35. Hope, P. S., Gibson, A. G., Ward, I. M.: J. Polym. Sci. Polym. Phys. Ed. *18,* 1243 (1980)
36. Kolbeck, A. G., Uhlmann, D. R.: J. Polym. Sci., Polym. Phys. Ed. *15,* 27 (1977)
37. Miller, R. L.: Polymer Handbook (eds.) Brandrup, J., Immergut, E. H., p. III-1-50, New York, Wiley 1975
38. Newman, S.: J. Polym. Sci. *47,* 111 (1960)
39. Williams, T.: J. Mater. Sci. *8,* 59 (1973)
40. Enns, J. B., Simha, R.: J. Macromol. Sci., Phys. Ed. *B 13,* 11 (1977)
41. Mead, W. T. et al.: Macromolecules, *12,* 473 (1979)
42. Shimada, T., Porter, R. S.: Polymer *22,* 1124 (1981)
43. Maruyama, S., Imada, K., Takayanagi, M.: Intern. J. Polym. Mater. *2,* 125 (1973)
44. Yokouchi, M. et al.: Macromolecules *9,* 266 (1976)
45. Ward, I. M.: Mechanical Properties of Polymers, p. 181, London, Wiley 1971
46. Gibson, A. G., Ward, I. M.: J. Polym. Sci., Polym. Phys. Ed. *16,* 2015 (1978)
47. Gibson, A. G., Davies, G. R., Ward, I. M.: Polymer *19,* 683 (1978)
48. Modena, M., Garbuglio, C., Ragazzini, M.: J. Polym. Sci., Polym. Lett. Ed. *10,* 153 (1972)
49. Cella, R. J.: J. Polym. Sci., Polym. Symp. *42,* 727 (1973)
50. Hoeschele, G. K., Witsiepe, W. K.: Angew. Makromol. Chem. 51. Hoeschele, G. K.: Chimia *28,* 544 (1974)
51. Hoeschele, G. K.: Chimia *28,* 544 (1974)
52. Wegner, G. et al.: Angew. Makromol. Chem. *74,* 295 (1978)
53. Yokoushi, M. et al.: Macromolecules *9,* 266 (1976)
54. Stein, R. S., Norris, F. H.: J. Polym. Sci. *21,* 381 (1956)
55. Bandara, U.: Diploma-Thesis, Freiburg 1979
56. McCrum, N. G., Read, B. E., Williams, G.: Anelastic and Dielectric Effects in Polymeric Solids, p. 515, London – New York – Sydney, John Wiley & Sons 1967
57. Farrow, G., McIntosh, J., Ward, I. M.: Makromol. Chem. *38,* 147 (1960)
58. McCrum, N. G., Read, B. E., Williams, G.: Anelastic and Dielectric Effects in Polymeric Solids, p. 564, London – New York – Sydney, John Wiley & Sons 1967

Received February 17, 1982
H.-J. Cantow (editor)

Author Index Volumes 1–47